Carbon Capture and Storage

Carbon Capture and Storage

Special Issue Editor

José Carlos Magalhães Pires

MDPI • Basel • Beijing • Wuhan • Barcelona • Belgrade

Special Issue Editor
José Carlos Magalhães Pires
University of Porto
Portugal

Editorial Office
MDPI
St. Alban-Anlage 66
4052 Basel, Switzerland

This is a reprint of articles from the Special Issue published online in the open access journal *Energies* (ISSN 1996-1073) from 2018 to 2019 (available at: https://www.mdpi.com/journal/energies/special_issues/Carbon_Capture_Storage)

For citation purposes, cite each article independently as indicated on the article page online and as indicated below:

LastName, A.A.; LastName, B.B.; LastName, C.C. Article Title. *Journal Name* **Year**, *Article Number*, Page Range.

ISBN 978-3-03921-399-3 (Pbk)
ISBN 978-3-03921-400-6 (PDF)

Contents

About the Special Issue Editor

José Carlos Magalhães Pires graduated in Chemical Engineering from the Faculty of Engineering of Porto University (FEUP) in 2004 with a final classification of 16. He worked in two chemical companies (SunChemical and Tagol) and then started his PhD in Environmental Engineering in 2006 at FEUP. From 2010 to 2016, he was a postdoctoral researcher in the area of environmental applications of microalgal cultures, focusing on the integration of processes to reduce operational costs and environmental impact. More recently, he was an Invited Assistant Professor (part-time) at FEUP and made a teaching contribution in the Faculty of Engineering of Agostinho Neto University (FEUAN) in Luanda, Angola. His teaching activity comprises several multidisciplinary topics, such as project engineering, chemical fundamentals, and air pollution, among others. JCM Pires received the best student prize from Fundação Engenheiro António de Almeida in 2005. Due to his scientific outputs of 2017, he received the Scientific Recognition Award at the 2019 FEUP day. JCM Pires has published 55 papers (22 as the first author, two as the single author, 26 as the corresponding author, and 14 as the senior/last author; 27 papers were published independently from his PhD supervisor) in international peer-reviewed journals (SCOPUS h-index = 20). JCM Pires has published three books and 17 book chapters, and his research work has been discussed in 40 international meetings. Since completing his PhD degree, JCM Pires has been a reviewer of 238 papers. He is the principal investigator of two research projects. JCM Pires has completed the supervision of one PhD student, seven Master's students, and two research fellows. Concerning established international collaborations, he worked in the Institute of Chemical Technology (Prague) for 5 months. He also contacted and established collaborations with researchers in Portugal (LSRE/LCM, CEFT, CIIMAR, and REQUIMTE) and abroad (Univ. of Cambridge, UK; Univ. of Rio de Janeiro and Univ. of Mato Grosso do Sul, Brazil; and Univ. Sevilla, JRC, Spain; among others).

Preface to "Carbon Capture and Storage"

Carbon dioxide is the most important greenhouse gas (GHG) due to the dependence of world economies on fossil fuels as an energy source, increasing the levels of CO_2 emissions in the atmosphere. This phenomenon is associated with several negative environmental impacts, including climate change and ocean acidification, among others. To mitigate this environmental problem, three options have been proposed: (i) application of carbon capture and storage (CCS) to reduce CO_2 emissions; (ii) improvement of energy efficiency; and (iii) the use of noncarbon energy resources (e.g., biomass, solar, and wind energy). The high costs of renewable energies along with the abundance and availability of fossil fuels delay the introduction of these environmentally friendly energy forms. There are also some barriers to changing the technological systems, which are designed around fossil fuel energy. Thus, in the coming years, energy will continue to be obtained mainly from fossil sources. Given this, CCS is the favored technology for stabilizing atmospheric CO_2 concentrations. It involves capturing CO_2 at the point of generation, compressing it to a supercritical fluid, and then sequestering it.

The CCS methodologies comprise three steps: CO_2 capture, CO_2 transportation, and CO_2 storage. CO_2 is captured at fixed point sources, such as power plants and cement manufacturing facilities, and different methods are studied with this aim. The most common are absorption, adsorption, separation by membranes, and cryogenic separation. Then, the captured gas mixture is compressed into a liquid and supercritical fluid to be transported by pipeline or ship to the place where it will be stored. The CO_2 storage options comprise geological storage, ocean storage, and mineralization. In essence, CCS keeps CO_2 out of the atmosphere by capturing it from exhaust gas and injecting it into deep reservoirs that can contain fluids for thousands of years. CCS is an important technological option because it allows societies to maintain their existing carbon-based infrastructure while minimizing the effects of CO_2 on Earth's climate system. However, this technology is still under development. This current book aims to evaluate different perspectives concerning CCS methodologies.

José Carlos Magalhães Pires
Special Issue Editor

Review

An Overview of the Portuguese Energy Sector and Perspectives for Power-to-Gas Implementation

Carlos V. Miguel *, Adélio Mendes and Luís M. Madeira

LEPABE—Laboratory for Process Engineering, Environment, Biotechnology and Energy, Faculty of Engineering, University of Porto, Rua Dr. Roberto Frias, 4200-465 Porto, Portugal; mendes@fe.up.pt (A.M.); mmadeira@fe.up.pt (L.M.M.)
* Correspondence: cvmiguel@fe.up.pt; Tel.: +351-22-508-1519

Received: 12 October 2018; Accepted: 17 November 2018; Published: 23 November 2018

Abstract: Energy policies established in 2005 have made Portugal one of the top renewable power producers in Europe, in relative terms. Indeed, the country energy dependence decreased since 2005, although remaining above EU-19 and EU-28 countries in 2015 (77.4% vs. 62.4% vs. 54.0%, respectively). Data collected from governmental, statistical, and companies' reports and research articles shows that renewables and natural gas assumed a growing importance in the Portuguese energy mix along time, while oil followed an opposite trend. Recently, the country remarkably achieved a full 70-h period in which the mainland power consumed relied exclusively on renewable electricity and has several moments where power production exceeds demand. Currently, the main option for storing those surpluses relies on pumped hydro storage plants or exportation, while other storage alternatives, like Power-to-Gas (PtG), are not under deep debate, eventually due to a lack of information and awareness. Hence, this work aims to provide an overview of the Portuguese energy sector in the 2005–2015 decade, highlighting the country's effort towards renewable energy deployment that, together with geographic advantages, upholds PtG as a promising alternative for storing the country's renewable electricity surpluses.

Keywords: CO_2 capture and utilization; energy dependence; power-to-methane; synthetic natural gas; renewable power; fossil fuels

---

1. Introduction

The Renewable Energy Roadmap 21 settles for 2020 and for the whole European Union a share of energy from renewable sources of 20% [1]. Some countries, such as Portugal, have already reached or surpassed such a target [2,3]; in fact, the current energy situation in the country has significantly changed in the last decade, when renewable energy deployment strategies were still under debate [4]. Portugal was the fourth country of the European Union with a higher incorporation of renewable electricity in 2015 (44.6%) after Denmark (50.2%), Austria (62.6%), and Sweden (72.1%) [3]. The Portuguese renewable annual electricity production has increased almost fourfold since 2005 and reached 33.3 TWh in 2016, relying mostly on hydro (16.9 TWh) and wind (12.5 TWh) sources, together representing 88% of the total renewable power production [3].

In Portugal, an annual surplus of renewable power production in the range of 800–1200 GWh is estimated for 2020 [5,6]. As renewable power relevance increases within the energy sector, developing a way to efficiently and economically store its surpluses in periods of low demand becomes an urgent problem to be tackled [7]. Among the systems available or under development for such a purpose (pumped hydroelectric storage, compressed air energy storage, electrochemical and flow batteries) [8], power-to-gas technologies (PtG) are receiving increased attention, particularly in Europe [9–11], and a storage potential of at least 500 GWh has been foreseen in Portugal [6]. One PtG option could be to use the surplus electricity for H_2O electrolysis to obtain H_2 (PtH), but its storage remains a challenge and lacks a dedicated infrastructure for its distribution [2]. Another way is to use that "green" H_2 and blend it in natural gas, but only up to 10% without major effect in the gas grid and end-use equipment, or further convert it to methane (PtM), also called substitute/synthetic natural gas (SNG), through the Sabatier reaction (Equation (1)) [9]. Methane is far simpler to store and transport than pure H_2 using the well-established natural gas infrastructure and therefore enabling the connection between the power and natural gas grids [12,13].

$$CO_2 + 4\,H_2 \rightleftharpoons CH_4 + 2\,H_2O \;\; \Delta H_{298\,K} = -165\;kJ \cdot mol^{-1} \tag{1}$$

Synthetic natural gas can be later reconverted to electricity in periods of high demand or used as feedstock or fuel. Thus, SNG can be seen as a secure and efficient supply of renewable energy, while simultaneously reducing the dependence on (imported) fossil fuels and supporting the transition towards a low-carbon economy [14–16].

Bailera et al. [10] reported the existence of 43 PtG projects worldwide taking place in 11 countries, with most initiatives occurring in Germany (16 projects), Denmark (7 projects), and Switzerland (6 projects) as a result of strong governmental support. In the review by Quarton and Samsatli [13], these results were updated, with Germany standing out among other countries with 45 projects, either finished, planned, operating, or under construction. The main drivers towards PtG in Germany are the existence of geographic advantages for PtG implementation, like the availability of enough suitable underground gas storage capacity and a sufficient gas network development for gas distribution [11,17], as well as the country targets to increase its power generation with origin in renewable sources from 32% (in 2015) to 50% and 80% in 2030 and 2050, respectively [18]. In Portugal, despite being a pioneering country regarding the adoption and massive diffusion of wind power parks across its territory, the first national research project in the country dedicated to the topic was launched in mid-2018 [19], dealing with the development of a cyclic sorption-reaction process for simultaneous CO_2 capture and conversion to methane to be coupled in PtG applications (cf. Figure 1).

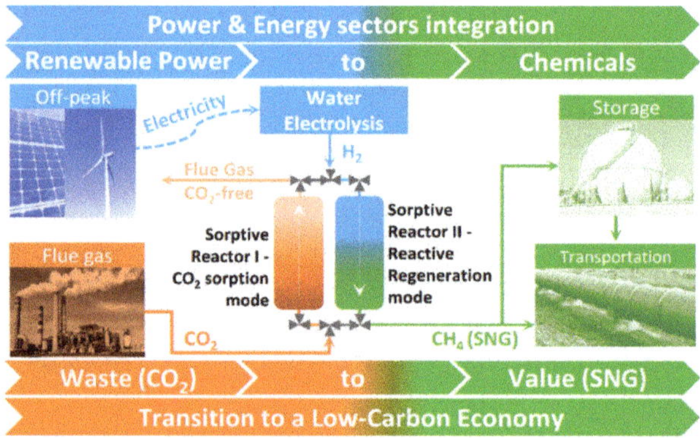

Figure 1. Power-to-gas concept: system boundaries with a cyclic sorption-reaction process for CO_2 capture and conversion/utilization. Reprinted from Chemical Engineering Journal, 322, C.V. Miguel, M.A. Soria, A. Mendes, L.M. Madeira, A sorptive reactor for CO_2 capture and conversion to renewable methane, 590–602, Copyright (2017), with permission from Elsevier.

There are few studies concerning the assessment of power-to-gas implementation potential in Portugal. The first work was by Heymann and Bessa [6], who estimated the cost of PtG products in the country as a function of the distance to wind power parks and gas storage facilities. The levelized cost of energy when considering SNG as a final product ranged between 0.05–0.10 €/kWh. Recently, Carneiro et al. [20] presented the opportunities for large-scale energy storage in geological formations in mainland Portugal.

While PtG demonstration activities are growing fast, particularly in Europe, the current situation in Portugal supports the findings by Bento and Fontes [21] that, typically, Portugal has an average adoption of energy-related technologies lag of one to two decades relative to "core" countries (i.e., energy technology developers/leaders, generally from the OECD-Organization for Economic Co-operation and Development) [21]. Hence, the present work aims to contribute to the current state-of-art by providing a background image of the Portuguese energy sector (Section 2), presenting the main facts and figures, such as the country energy dependence evolution with time (Section 2.1), the consumption of fossil fuels (Section 2.2), and renewable power production (Section 2.3). Afterwards, in Section 3, requirements for power-to-gas implementation are described, namely the availability of renewable power surpluses (Section 3.1), carbon dioxide sources for the methanation (Section 3.2), and access to the natural gas grid for SNG storage and distribution (Section 3.3). In Section 3.4, needs for future research are identified and, finally, in Section 4, the main conclusions and the most important steps that all interested parties should take to raise awareness regarding deployment of PtG in Portugal are presented. Figure 2 shows a diagram presenting the approach adopted in this work.

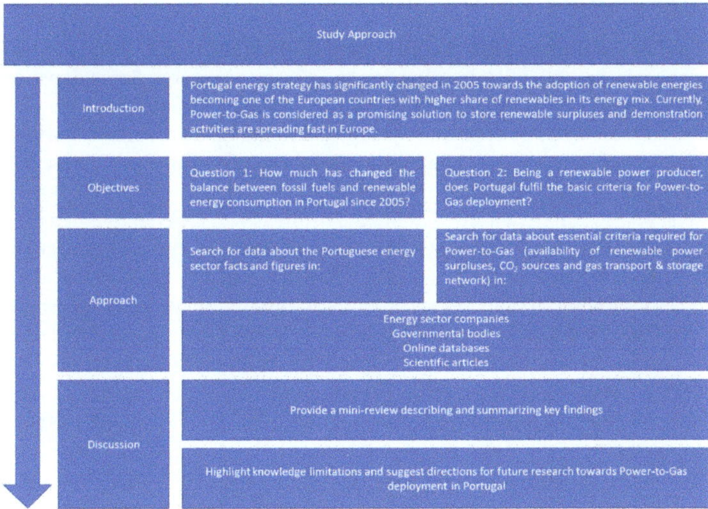

Figure 2. Diagram presenting the study approach adopted in this work.

2. Overview of the Portuguese Energy Sector

2.1. Energy Dependence

The energy dependence (*ED*) is a parameter that characterizes the extent to which an economy relies upon imports to meet its energy needs. The indicator is calculated as net imports of primary energy (i.e., (*IMP*) importations minus exportations (*EXP*)) divided by the sum of gross inland energy consumption (*GIC*) plus international maritime bunkers (*IMB*) (cf. Equation (2)) [22].

$$ED(\%) = \frac{IMP - EXP}{GIC + IMB} \times 100 \tag{2}$$

The Portuguese energy dependence and the dependence of the Euro-economic area (EU-19) and European Union countries (EU-28) are shown in Figure 3 for comparison.

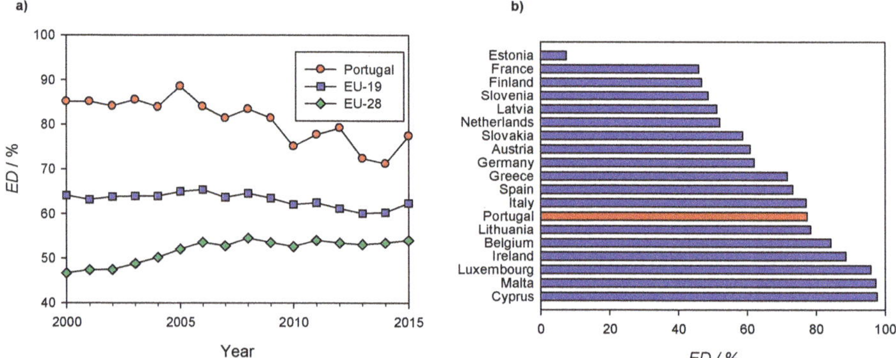

Figure 3. Portuguese energy dependence (*ED*): (**a**) Along recent years and (**b**) comparison with EU-19 countries in 2015. Data taken from Eurostat online database [22].

Portugal had the seventh highest energy dependence among the EU-19 and EU-28 countries in 2015 (cf. Figure 3b). None of the EU-19 countries had a negative energy dependence (cf. Figure 3b), all depending on primary energy imports to satisfy their energetic needs.

The normalized consumption of primary energy (CPE) per type of source in Portugal is shown in Figure 4a, for the period of 2000–2015. The country situation is compared with those from EU-19 and EU-28 group countries for the year, 2015, in Figure 4b.

The National Energy Strategy, approved in 2005 by the Portuguese Government, settled strategic policies, such as the energy market liberalization, the promotion of energy from renewable sources, and of technologies with improved efficiencies [23]. As a result, the oil share remarkably declined (i.e., 14.6%) in the following decade, replaced by natural gas and energy from renewable sources, whose values increased by 4% and 9%, respectively, while the coal share practically remained constant in the same period (a rise of only 1.6%) (cf. Figure 4a). Still, fossil fuels represented 78% of the consumed primary energy in 2015, a value slightly above EU-19 (72%) and EU-28 (73%) group countries, whose patterns are nearly identical (cf. Figure 4b). The remaining primary energy was exclusively based on renewable sources (22%), making Portugal the fifth country with the highest share of energy from renewables amongst the EU-28 countries [24]. The weight of renewables becomes more significant when considering primary energy consumption exclusively for power production purposes. In fact, 45% of the electricity produced in 2015 was obtained from renewable sources [3]. Nuclear has almost the same weight as the energy from renewable sources (ca. 13–15%) in EU-19 and EU-28 groups, although it is absent in some members, like Portugal.

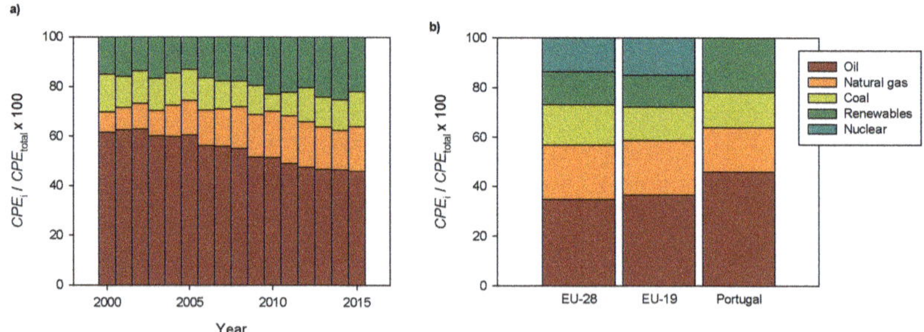

Figure 4. Normalized primary energy consumption per type of source in Portugal: (**a**) From 2000 to 2015 and (**b**) comparison with EU groups in 2015. Authors calculations based on data taken from Eurostat online database [22] and excluding the negligible contribution of non-renewable waste sources.

The last 10 years up to the present regarding fossil fuels and energy from renewable sources contributions to the Portuguese energy sector is presented in Sections 2.2 and 2.3, respectively, with an emphasis on renewable power production, as it is one of the main building blocks of power-to-gas technologies.

2.2. Energy from Fossil Fuels

2.2.1. Oil

Up to now, Portugal does not have indigenous oil reserves with economic viability, although regular onshore and offshore exploration activities have been carried out since 1940. Therefore, all oil consumed by the country is imported. Table 1 lists the top five supplier countries from 2014 to 2016. In the listed years, Portugal imported oil from 13–15 countries and the top five oil suppliers were responsible for around 66–76% of the total imported oil. Angola was the major oil supplier with a contribution of ca. 25%. Diversification of oil suppliers along the years has contributed to assure reliable and secure access to fossil energy resources [25].

Table 1. Top five oil suppliers to Portugal and their corresponding share (based on data taken from [26]).

Top-5	2014	2015	2016
1st	Angola (26.1%)	Angola (22.9%)	Angola (24.9%)
2nd	Saudi Arabia (12.6%)	Saudi Arabia (14.2%)	Russia (19.7%)
3rd	Algeria (9.9%)	Kazakhstan (10.6%)	Azerbaijan (11.1%)
4th	Kazakhstan (9.7%)	Algeria (9.5%)	Saudi Arabia (10.8%)
5th	Azerbaijan (9.2%)	Azerbaijan (9.0%)	Kazakhstan (9.3%)
Imp. oil (10^6 ton)	7.5 (out of 11.17)	9.1 (out of 13.73)	10.7 (out of 14.09)
Nr. oil suppliers	14	15	13

Figure 5 shows the final consumption of oil by activity sector. The transportation sector is responsible for the largest share (ca. 75–79%). Oil consumption declined in all sectors for the five year period, except Agriculture/Forestry, which remained practically constant. Within the Portuguese industry sector, the non-metallic minerals industries (e.g., cement and glass) were by far the activities with higher oil consumption (i.e., 50–60% of the oil consumed by the industry sector).

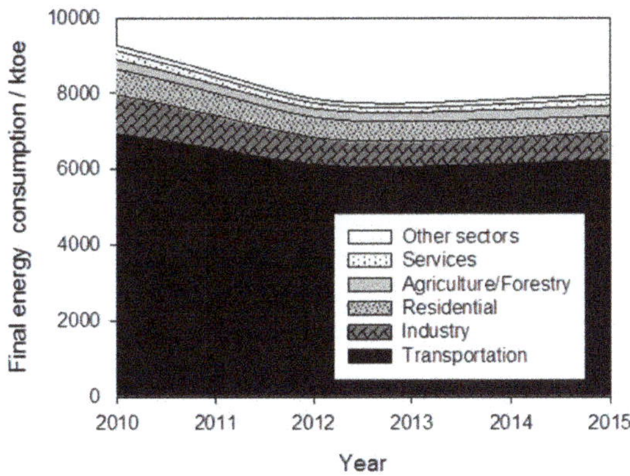

Figure 5. Evolution of oil products' final energy consumption in Portugal by activity sector from 2010 to 2015 (data taken from [27]).

2.2.2. Coal

After national coal production ceased in 1994, Portugal dependence on imported coal to secure its energy needs increased. Portugal imported 4.5 millions of tonnes of coal from Colombia (88.1%), the United States (6.6%), South Africa (3.5%), and Ukraine (1.8%) in 2014 [24]. Imported coal is of the bituminous type, being used essentially for electricity generation in two coal-fired power plants located in Sines (1250 MW) and Pego (620 MW). These plants act as a backup system, guaranteeing that power demand is fulfilled in periods of low renewable power production. Coal consumption is particularly dependent on the hydrological conditions, namely when hydropower output is lower during drought periods. Coal is also consumed by end-users from the industry sector, namely by the iron and steel industries, in chemical/petrochemical plants, and by the non-metallic minerals industries (cf. Figure 6). Nevertheless, the amount of coal used by these end-users is negligible when compared to the quantity used for electricity production (e.g., 12 ktoe vs. 3246 ktoe, respectively). Still, coal consumption by the non-metallic minerals and chemical sectors has decreased considerably from 2010 to 2015 (cf. Figure 6). The amount of coal consumed in 2014 and 2015 by the non-metallic minerals, iron and steel, and chemical/petrochemical sub-sectors was similar (ca. 4 ktoe/each) (Figure 6).

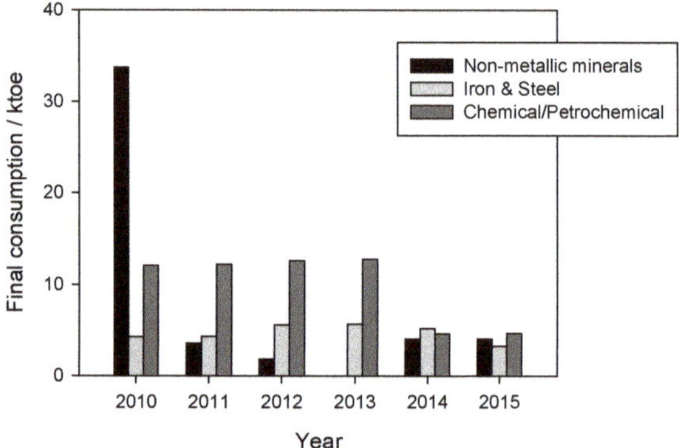

Figure 6. Coal consumption by the industry sub-sectors from 2010 to 2015 [27].

2.2.3. Natural Gas

Portugal has no natural gas resources. Table 2 shows natural gas import origins and corresponding volumes for the years, 2014 and 2015.

Table 2. Natural gas imports (10^6 m^3) in the years 2014 and 2015 [28].

Delivery Type	Origin	2014	2015
Pipeline (Natural gas)		**2736**	**3002**
	Algeria	2196	2111
	Spain	535	891
	Not specified	5	0
Ships (Liquefied natural gas-LNG)		**1523**	**1776**
	Algeria	102	210
	Qatar	687	224
	Nigeria	352	1166
	Norway	80	80
	Spain [1]	6	7
	Trinidad and Tobago	223	89
	Not specified	73	0
TOTAL		**4259**	**4778**

[1] LNG imported using tanker trucks.

Around 64% of the supplies were received through a pipeline, while the remaining part, liquefied, was transported to Portugal in ships that unload at the Sines terminal, on the southern part of the country. Only a negligible quantity ($6–7 \times 10^6$ m^3) was imported using tanker trucks exclusively from Spain. The most important supplier is Algeria, with a share ranging between 45–52%, while Qatar and Nigeria were the major suppliers of liquefied natural gas (LNG).

The final energy consumption of natural gas by activity sector is shown in Figure 7. The industry sector accounts for the largest amount of natural gas consumption (67–74%), followed by the residential (16–19%) and services (13–14%) sectors. The use of natural gas in the agriculture/forest and transportation sectors is negligible and both sectors represent only ca. 1% of the total consumption. Energy for transportation purposes is assured predominantly by oil products (as shown in Section 2.2.1), with natural gas playing a negligible role; for instance, the quantity of oil and natural gas consumed in 2015 for transportation was 6245 ktoe vs. 13 ktoe, respectively.

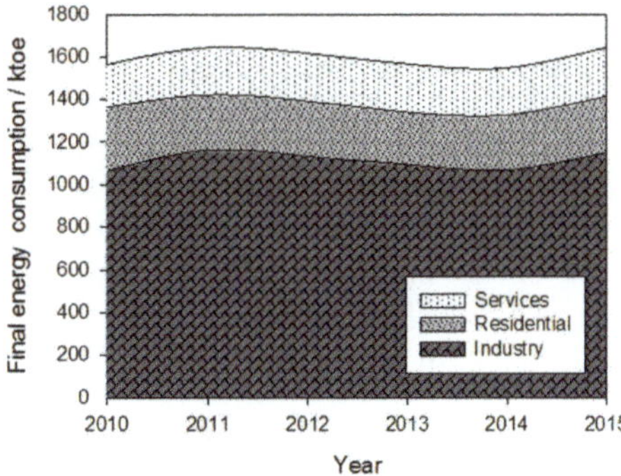

Figure 7. Natural gas for final energy consumption in Portugal by activity sector from 2010 to 2015 (data taken from [27]).

Table 3 lists the natural gas consumption across the industry sectors during 2005, 2014, and 2015.

Table 3. Natural gas consumption (ktoe) by the Portuguese industry (data taken from [27]).

Industry	2005	2014	2015	Δ (2015/2005)
Paper, Pulp, and Print	38.1	90.2	111.6	2.93
Construction	5.8	13.1	14.5	2.48
Chemical and Petrochemical industry	64.4	142.1	152.7	2.37
Food and Tobacco	66.5	124.6	147.2	2.21
Non-ferrous metal industry	7.6	12.9	16.0	2.09
Machinery	21.3	32.6	36.0	1.69
Iron & steel industry	41.4	47.4	51.1	1.23
Textile and Leather	128.6	131.4	131.9	1.03
Non-metallic Minerals (e.g., cement)	516.6	426.3	441.6	0.85
Wood and Wood Products	9.7	9.0	7.8	0.81
Mining and Quarrying	6.3	5.1	4.6	0.74
Transport Equipment	28.8	14.9	16.6	0.58
Non-specified (Industry)	20.9	7.2	5.8	0.28
Total	**956.0**	**1056.8**	**1137.4**	**1.19**

The values listed in Table 3 show that the non-metallic minerals sector is the biggest consumer of natural gas. Table 3 also highlights the growing relevance of natural gas over time. Indeed, since 2005, the annual consumption increased in eight out of 13 industrial activities (see relative variation in the last column). Amongst them, the paper, pulp, and print, the chemical and petrochemical, and the food and tobacco (2.21) industries stand out given their absolute energy consumption and relative variation values, more than duplicating in all of them in only 10 years. Globally, natural gas consumption increased 19% in the 2005–2015 decade, which reflects its growing importance for the Portuguese industry sector.

2.3. Energy from Renewable Sources

The strategic effort to replace fossil fuels by energy from renewable sources has made Portugal one of Europe's leaders in this area [24]. Table 4 lists the amount (in ktoe) of energy from renewable sources produced in Portugal during the 2005–2015 decade.

Table 4. Portuguese annual production of energy from renewable sources (ktoe) from 2005 to 2015 [3].

Renewable Energy Type	2005	2007	2009	2011	2013	2015	Δ (2015/2005)
Biofuels	0	162	226	330	274	321	-
Electricity [1]	599	1265	1456	1872	2369	1927	3.2
Biomass [2]	2773	2891	3019	2571	2812	2781	1.0
Other renewables [3]	20	23	36	61	74	82	4.1
Total	**3392**	**4342**	**4737**	**4835**	**5530**	**5110**	**1.5**

[1] Includes the contribution of hydro, wind, photovoltaic, and geothermal power; [2] includes the contribution of biogas; [3] includes solar (for thermal purposes) and (low enthalpy) geothermal sources.

Since 2006, Portugal has produced biodiesel, which is incorporated almost completely in the conventional fossil diesel and only a small fraction (ca. 1%) is directly sold in the market. Soybean and, particularly, colza oils are the most used raw materials [29]. More than half of the energy from renewable sources produced in Portugal comes from biomass, although that share decreased from 82% to 54%, when comparing the values of 2005 and 2015. The amount of energy produced from biomass remained nearly constant along the 2005–2015 decade, while the production of electricity tripled, reaching a share of 38% of the total energy from renewable sources produced in 2015 (cf. Table 4). Electricity production values shown in Table 4 include contributions from hydro, wind, photovoltaic, and geothermal sources, and excludes contributions from biomass in thermoelectric and co-generation plants. Information regarding the present energy production status from biofuels and biomass can be found elsewhere (e.g., [24]).

The investment made on the different technologies for power production from renewable sources is highlighted through the analysis of the installed capacity (MW) values listed in Table 5. The most established renewable energy sources (RES) for electricity production in Portugal are hydro and wind, both totaling over 90% of the installed capacity. Biomass is the third RES with a higher installed capacity, followed closely (in recent years) by photovoltaic, which remarkably increased from 3 MW to 451 MW in the 2005–2015 decade. During this period, the wind energy installed capacity increased almost 400% and was by far the type of RES with the highest absolute variation (i.e., 3971 MW).

Table 5. Renewable energy sources' installed capacity (MW) in Portugal for electricity production and corresponding variation in the 2005–2015 decade [3,30].

RES	2005	2007	2009	2011	2013	2015	Δ (2015/2005)
Geothermal	18	29	29	29	29	29	1.6
Photovoltaic	3	15	110	175	299	451	150.3
Biomass	429	449	518	712	718	726	1.7
Wind	1063	1699	3564	4378	4731	5034	4.7
Hydro	4816	4853	4883	5330	5533	6053	1.3
Total	**6329**	**7045**	**9104**	**10,624**	**11,310**	**12,293**	**1.9**

Among biomass, it should be mentioned the evolution of biogas production, whose installed capacity increased from 8 MW (in 2005) to 85 MW (in 2015), while the capacity for energy generation from urban solid wastes only increased 3 MW, reaching a total capacity of 89 MW in 2015.

The exploitation of the installed capacity for power production from the different RES is provided in Table 6, which shows that, globally, the power production tripled in the 2005–2015 decade. In the following sections, the status of each RES for producing electricity is addressed.

Table 6. Annual renewable power production (GWh) in Portugal and corresponding variation in the 2005–2015 decade [3,30].

RES	2005	2007	2009	2011	2013	2015	Δ (2015/2005)
Geothermal	71	201	184	210	197	204	2.9
Photovoltaic	3	24	160	282	479	799	266.3
Biomass	1651	1883	2086	2924	3052	3104	1.9
Wind	1773	4036	7577	9162	12,015	11,608	6.5
Hydro	5118	10,449	9009	12,114	14,868	9800	1.9
Total	**8616**	**16,593**	**19,016**	**24,692**	**30,610**	**25,514**	**3.0**

2.3.1. Geothermal

Among the RES, high temperature geothermal resources are confined to the Azores archipelago where this kind of energy plays an important role. Two geothermal power plants in operation at S. Miguel island, corresponding to a global installed capacity of 23 MW, are responsible for the production of 42% of the consumed electricity (i.e., around 22% of the archipelago total demand). Plans to increase the installed capacity up to 28.5 MW until 2019 have been reported [31]. The International Energy Agency (IEA) reported that enhanced geothermal systems technology, which uses thermal energy from high-temperature rocks (dry rocks) located at great depths, may be suitable to explore the potential geothermal resources in the mainland and be tested in the future [24]. Still, Portugal is the fifth country among IEA-29 members with the highest share of geothermal energy used for power production [24].

2.3.2. Photovoltaic

Power production in photovoltaic plants was negligible in 2005 and reached 799 GWh in 2015, being the RES with the highest relative variation in the 2005–2015 decade (cf. Table 6). Portugal has the best yearly solar irradiance in Europe after Cyprus, particularly in the Alentejo region, in the southern part of the territory, where the country has a current installed capacity of 162 MW (out of a total of 467 MW) [3,32]. The photovoltaic plant located in Moura is the largest in the country comprising an installed capacity of 46 MW. It is expected that solar energy will play an important role in decentralised power production, and a mini-generation programme created in 2011 has a target to install approximately 250 MW of new capacity by 2020 [24]. Before 2011, the lack of specific regulations for mini-generation systems limited photovoltaic diffusion as the feed-in tariffs settled in 2007 by the Decree-Law No. 225/2007 have not been listed explicitly, being calculated monthly for each system based on avoided costs, which leads to administrative difficulties as well as low transparency [33].

2.3.3. Biomass

The most common biomass resources available in Portugal are wood residues, animal waste, and municipal solid waste [32]. It was estimated that the country's total biomass potential is 42.5 TWh/year, with municipal solid wastes as the main resource (17.0 TWh/year) [32]. It has been reported that the use of municipal solid wastes, animal manure, and wastewaters are still underexploited [32]. In 2015, 586 GWh of power was generated from biogas and urban solid wastes (ca. 294 GWh each), together representing 19% of the total power produced from biomass (i.e., 3.10 TWh).

Power production from biomass is more developed in the center region of the country, representing 62% of the total power produced from biomass in 2015.

2.3.4. Hydropower

Hydropower production takes place in 184 hydropower plants, considering both large ($\geq$10 MW) and small plants (<10 MW) [34]. Portugal mainland's most important river basins are: Lima, Cávado, Mondego, Tejo, Guadiana, and, particularly, Douro, which is responsible for more than half of the hydropower generated in the country [3]. Portugal has storage, run-of-the-river, and pumped hydro storage type hydropower plants. Storage plants accumulate large quantities of water that can be used on the driest months, while run-of-rivers may include a small storage capacity, and turbines operate depending on the river's flow. Pumped hydro storage plants are conventional plants that were modified to include a system for pumping water from a lower elevation reservoir to a higher elevation. Generally, low cost power for running the pumps is provided by off-peak electricity generated from renewable energy sources, allowing the storage of that energy in the form of gravitational potential energy [35]. Portugal has a pumped hydro storage installed capacity of 2.44 GW [36].

Table 7 lists the hydropower generation per type of plant and river basin in the year of 2015. To the authors' knowledge, the values of power production through pumped hydro storage plants per river basin are not publicly available, but a global production of 1.16 TWh in 2015 was reported by Redes Energéticas Nacionais (REN, Lisboa, Portugal) [36].

Table 7. Hydropower generation (GWh) by type of plant and river basin in 2015 [3].

River Basin	Storage	Run-of-River	Total	%
Lima	484	5	489	5.0
Cávado	1180	29	1209	12.3
Douro	366	5422	5788	59.1
Mondego	322	88	410	4.2
Tejo	415	320	735	7.5
Guadiana	812	0	812	8.3
Others	0	355	355	3.6
Total	**3579**	**6219**	**9798**	**100**

Hydropower production in 2015 was affected by the hydrological conditions, and the amount of generated power was considerably lower than typical values found, representing only 60% of the power produced in 2014. Still, Table 7 highlights the importance of the global hydropower production of storage and run-of-river plants located in Cávado and Douro river basins, respectively.

2.3.5. Wind

Portugal had 255 wind parks with 2604 turbines in operation in 2015, corresponding to a total installed capacity of 5034 MW [3]. Figure 8 shows how the installed capacity, wind power production, and annual equivalent hours at full capacity (HFC)—ratio between the generated output (MWh) and the installed capacity (MW)—were distributed countrywide in 2015.

Figure 8 highlights that wind power production is massively obtained in the Center and North regions, together representing 87% of the overall production [3]. Globally, wind power was generated in 2305 equivalent hours at full capacity, with the North, Azores, and Madeira the only regions with an HFC lower than the global.

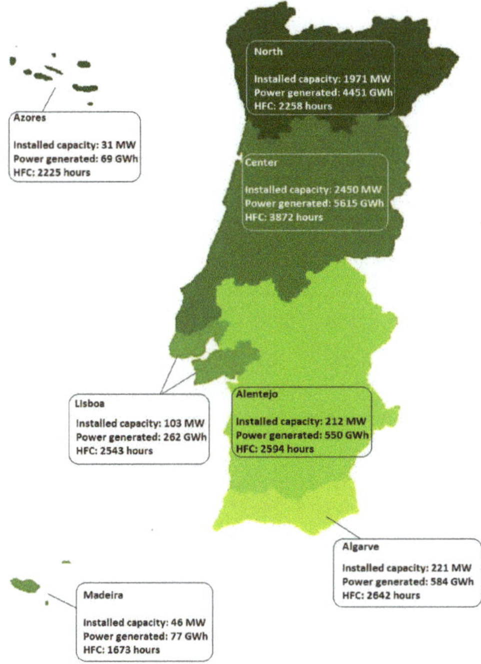

Figure 8. Wind power installed capacity, generation, and annual equivalent hours at full capacity (HFC) for the year 2015 (data taken from [3]).

The performance of wind turbines can also be expressed in terms of a capacity factor, where the number of equivalent hours at full load is normalized by the number of hours available in a given period, allowing assessment of the amount of power that was produced with respect to the maximum possible. Table 8 lists the capacity factor of each region for an ideal availability of 8760 h (i.e., hours available in one year without excluding shutdown periods (e.g., for maintenance)).

Table 8. Annual wind power capacity factor obtained in different regions of Portugal in 2015 (calculations based on data taken from [3]).

Region	Capacity Factor
Center	0.44
Algarve	0.30
Alentejo	0.30
Lisbon	0.29
North	0.26
Azores	0.25
Madeira	0.19

The annual capacity factor ranged from 0.19 (Madeira) up to 0.44 (Center), while the remaining regions had a value between 0.25 and 0.30. However, it is noteworthy to mention that the capacity factor can be considerably different depending on the period considered (year, month, etc.). Silva et al. [37] used historic wind power generation time series (up to five years) and reported that the capacity factor in Portugal can be 1.5 times higher in winter than in summer. The analysis on an hourly basis also showed that wind electricity generation is greater during base-load and off-peak periods. Unfortunately, the amount of wind power lost during these periods was not reported.

3. Perspectives for Power-to-Gas

3.1. Surplus Renewable Power

The IEA reported that instantaneous and daily renewable electricity output in Portugal regularly exceeds national demand and that the surpluses are either used in pumped hydro storage plants or exported [24]. Table 9 lists the amount of power that was consumed and produced in the country through pumped hydro storage in the period between 2010 and 2015.

Table 9. Pumped hydro storage power consumption and production (GWh) (calculations based on data taken from [27]).

Pumped Hydro Storage Power (GWh)	2010	2011	2012	2013	2014	2015	Δ (2015/2005)
Consumed	512	737	1331	1459	1081	1460	2.85
Produced	399	575	1038	1138	843	1139	2.85

Table 9 shows that the amount of surplus power consumed in pumped hydro storage plants almost triplicated in 2015 compared to the 2010 value; the storage round-trip efficiency is 78%. However, it should be recalled that pumped hydro storage plants represent huge capital investments and have been associated with several environmental impacts that are critical decisive factors [35,38]. Another option is to export the surpluses, in this case, to Spain. However, it has been reported that electricity from feed-in tariff supported technologies (like wind) is exported, and Portugal provides cheap electricity to Spain partially supported by the Portuguese Electricity System [39]. The feed-in tariff, a mechanism used by countries to foster the use of energy from renewable sources, is always paid to power producers independently of the power generated being used in Portugal or outside. Moreover, the tariff was guaranteed to power producers for a 20-year period, the longest reported among several European countries [39]. Therefore, it was recommended that exports should be reduced at high-generation moments, releasing the condition of feeding all renewables to the grid, and allowing for the spill of wind generation and/or investment in storage technologies [39].

Pumped hydro storage plants have been, to date, the main approach adopted in Portugal for the storage of excess renewable power. Decentralized units would, however, contribute to increase the country's storage capacity and foster energy transition as the substantial investment costs, appropriate geography, and inherent environmental impacts limit the extension of pumped hydro storage. To this end, power-to-gas technologies, particularly power-to-methane, could be selected based on the increasing consumption of natural gas (as discussed in Section 2.2.3), offering the possibility to integrate the power and gas grids. The renewable power is thus chemically stored as methane, which can be used on-site, injected into the natural gas grid, or stored in dedicated reservoirs, such as in salt caverns or in LNG tanks after being compressed.

Power-to-methane (PtM) applications in Portugal would benefit from the high quantity of installed wind power capacity within close distances of the gas infrastructure, with almost 60% of that capacity located less than 5 km to existing or future potential natural gas storage facilities, making Portugal a predestined country to implement PtM technologies, as long as adequate and nearby CO_2 sources are also available [6].

3.2. CO_2 Sources and Availability

The data presented in this section refers exclusively to greenhouse gas (GHG) emissions of CO_2 and not to other GHG species and their corresponding CO_2 equivalents. Figure 9 illustrates the evolution of CO_2 emissions in the country from 2005 to 2015, using data taken from the GHG emissions inventory regularly performed by the Portuguese Environmental Agency. It shows that CO_2 emissions are essentially divided into two categories: the most relevant class is related to CO_2 generated from the combustion of fuels for energy production (ca. 90% of all CO_2 emitted in 2015) and the other to

CO_2 produced in industrial processes. It should be recalled that emissions from biomass combustion are excluded from the national emissions totals since released carbon had been, in fact, fixed from the atmosphere by the photosynthetic process and, when it is burnt, returns to the atmosphere and does not increase the atmospheric/biosphere CO_2 pool [40]. However, CO_2 emitted from biomass combustion accounted for 11.5 Mt in 2015 [40].

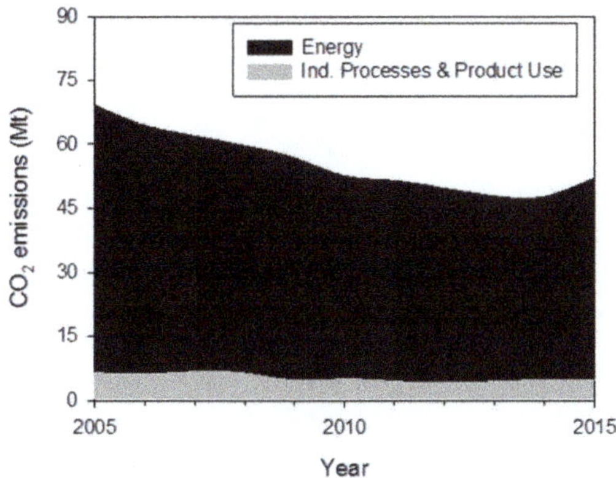

Figure 9. CO_2 emissions' evolution resulting from energy production and industrial processes and product use in Portugal for the period of 2005–2015 (adapted from [40]).

Regarding the evolution of the emissions history, Figure 9 shows that while CO_2 production from industrial processes remained almost constant in the reported period, the emissions due to energy production declined from 62.5 Mt to 42.9 Mt between 2005 and 2014 due to a combination of increased renewable power production and economic slowdown [24]. A slight increase of emissions was observed in 2015 (total of 52 Mt), which was a reflex of higher primary energy consumption for power production, namely of coal and natural gas. Additionally, final energy consumption also increased, particularly in road transport, natural gas, and electricity [40]. Figure 10 shows CO_2 emissions related to energy production in 2015 by type of sector and sub-sector.

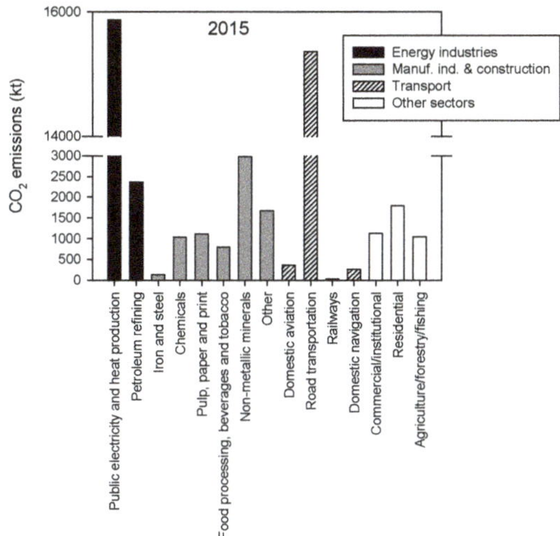

Figure 10. CO_2 emissions due to energy production by sector (2015) (adapted from [40]).

The amount of CO_2 emissions from public electricity and heat production and road transportation stand out amongst other sub-sectors (Figure 10). Together, these sub-sectors are responsible for 31 Mt of emitted CO_2, which corresponds to 60% of the national total. Afterwards, the non-metallic minerals sub-sector was responsible for the emission of 2981 kt of CO_2, being the principal emitter amongst the manufacturing industries and construction sector. The largest point emission sources for energy production considered were: 16 power plants, 2 oil refineries plants, 1 iron and steel industry, 1 petrochemical unit, 1 carbon black industrial plant, 8 paper pulp plants, and 6 cement plants. Besides energy production, the mineral industry also stands out in what concerns CO_2 emissions resulting from industrial processes (cf. Figure 11).

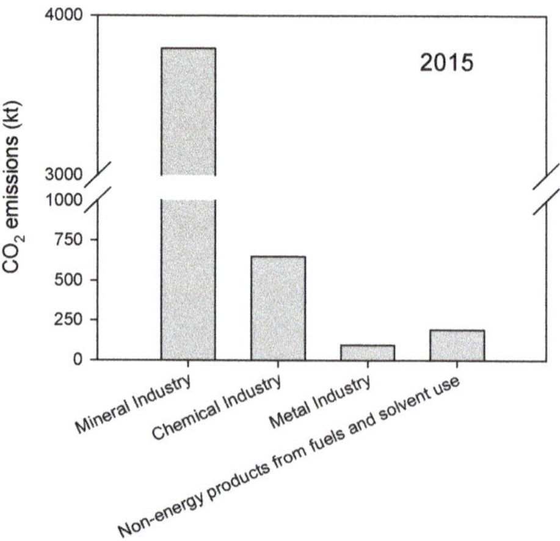

Figure 11. CO_2 emissions from industrial processes and product use (non-energy use) in 2015 (adapted from [40]).

The mineral industry sector was responsible for most of the CO_2 emissions in 2015 (3794 kt), with cement production being the most relevant activity with 2921 kt, followed by other processes, such as fertilizers (354 kt), lime (351 kt), and glass (167 kt) production (Figure 11). The second largest CO_2 emitter sector was the chemical Industry, namely by the petrochemical and carbon black production activity with 650 kt. The non-energy products from fuels and solvent use comprises emissions resulting from solvents, lubricants, and paraffin wax uses by several industries (e.g., plastics, wood products, rubber industry, and metalworking industry). CO_2 emissions associated to the metal industry come from secondary steel making.

From the analysis of Figures 10 and 11, it can be concluded that Portugal offers a wide variety of CO_2 sources, essentially diluted in flue gas streams, with a content between 5 vol. % (natural gas combustion) to 40 vol. % (e.g., cement) that can be selected to couple in power-to-methane applications, requiring, however, a previous CO_2 capture/purification stage to separate it from other contaminants. Criteria to select the best CO_2 source for power-to-methane applications include:

1. continuous access to CO_2 in a stream having low concentrations of severe poisons (e.g., H_2S and NO_x.);
2. proximity to the national natural gas grid for methane injection to avoid/minimize storage and transportation costs;
3. proximity to renewable electricity plants that will power the water electrolysis unit in periods where production exceeds demand, minimizing distribution losses;
4. interest on recycling the methane produced, for instance, if the selected site has a natural gas co-generation plant (i.e., displacement of fossil fuels consumption); and

5. interest on recycling the oxygen produced during water electrolysis to the process, which would further benefit the whole process from, at least, the economic point-of-view.

The CO_2 sources emitting more than 0.1 Mt/year in Portugal (data from 2007) and their geographic situation were identified by Carneiro et al. [41] and are illustrated in Figure 12.

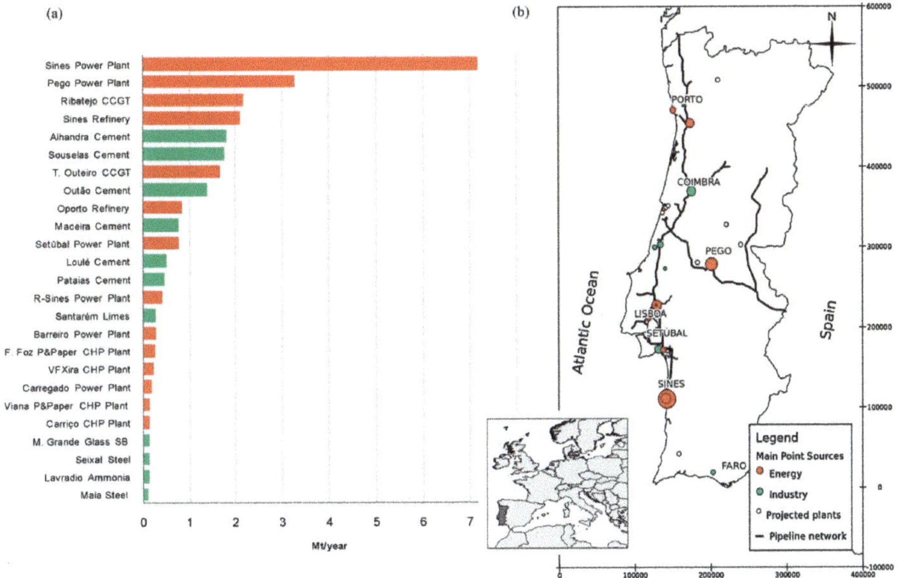

Figure 12. (**a**) CO_2 sources emitting more than 0.1 Mt/year in 2007 and (**b**) location of point sources and the natural gas network. Reprinted from International Journal of Greenhouse Gas Control, 5, J.F. Carneiro, D. Boavida, R. Silva, First assessment of sources and sinks for carbon capture and geological storage in Portugal, 538–548, Copyright (2011), with permission from Elsevier.

Figure 12 shows that all the main point sources are located close to the natural gas network, which, in addition to proximity to wind parks, as mentioned in the previous section, surely provides promising opportunities for PtG demonstration activities in the country, since CO_2 transport/storage practical difficulties and related cost uncertainties [11] might be avoided/minimized.

3.3. Natural Gas Grid

Portugal has a well-established natural gas storage, transportation, and distribution infrastructure (cf. Figure 13). The main pipeline goes next to the coast, from Sines until Valença do Minho (and onwards to Spain), where the main natural gas consumption points are located. It has several branches and two lines towards the interior of the country, one of which ends in Campo Maior and makes the connection with the Spanish pipeline in Badajoz (cf. Figure 13). This interconnection allows the country to receive up to 3.5×10^9 m^3/year of natural gas from Spain, whose origins are in Algeria, while the interconnection in the north (Valença do Minho/Tuy) has a lower capacity (0.8×10^9 m^3/year). Still, the highest entry capacity to the grid is provided by the LNG terminal in Sines (i.e., 5.3×10^9 m^3/year) [24]. Both interconnections with Spain are fully reversible and a third one in Bragança/Zamora (dashed line in Figure 13) is planned and identified by the European Commission as a project of common interest [24,28].

Figure 13. Portugal natural gas storage and transportation infrastructure (reprinted from [36] with permission from REN).

The national pipeline network has an extension of 1375 km with 202 pipeline stations [24]. Sines terminal receives LNG from large vessel ships with a capacity from 45×10^3 up to 216×10^3 m^3. These ships unload into three LNG storage tanks having a combined capacity of 390×10^3 m^3, corresponding to ca. 242×10^6 m^3 of natural gas [24,36]. The terminal is equipped with five vaporizers using sea water as thermal fluid to gasify LNG, which is further compressed to 78 bar and injected into the gas grid [24]. The terminal facilities also include a filling station that may load up to 4500 tanker trucks a year to distribute natural gas to locations not covered by the pipeline network [24].

Another fundamental element of the national natural gas grid is the combined underground storage capacity of 333×10^6 m^3 provided by six salt caverns located in Carriço. These caverns belong to the Monte Real salt structure of the Lusitanian basin, strategically placed in the middle of the main high-pressure pipeline (cf. Figure 13) [36,42]. The reasons for its construction were: (1) the storage of strategic reserves and (2) to balance supply and demand, namely due to seasonal and daily fluctuations, thus securing natural gas supply [42].

Carriço's underground storage facilities allow a gas injection and withdrawal of 110×10^3 Nm^3/h and 300×10^3 Nm^3/h, respectively. Before injection into the grid, the gas is filtered to remove solid and liquid particles, compressed, and dehydrated in a vertical absorber (the maximum final gas moisture content is 40 ppmv) [42].

The LNG terminal and Carriço salt caverns provide a total storage capacity of 575×10^6 m^3. Considering that consumption in 2016 was 4.6×10^9 m^3 [36], the existing combined capacity can stock the equivalent of the amount consumed by the country in 46 days. In the development plan of REN, the operator of the gas network, the construction of 25 new caverns in Carriço is forecasted to increase the storage capacity up to 1.25×10^9 m^3 [43], although these expansion plans were reported to be currently under review [24].

Additional underground storage capacity in the Portuguese territory was estimated by Nunes [43]. Several criteria were adopted to choose the best locations. The criteria included rejecting zones that were in a close distance to airports, roads, and houses, inside protected areas, far away from the sea and gas grid, or not in a plain field. Afterwards, three regions were elected: Nazaré, Caldas da Rainha, and Peniche. The study considered a similar cavern volume and distance among the caverns, like the Carriço facilities, and an underground storage potential of 1×10^9 m^3 was estimated. If the minimum distance to roads was limited to highways and railways, the storage potential reached 1.65×10^9 m^3 [43]. Hence, a potential storage capacity of 3.14×10^9 m^3 is envisaged, safeguarding 249 days of consumption (based on 2016 data). However, the preliminary assessment of the underground storage potential made by Nunes [43] should be complemented with the necessary environmental impact and economic studies. Recently, Carneiro et al. [20] screened priority sites for energy storage in geological formations using a geographic information system (GIS) and considered spatial, environmental, and social constraints, as well as the proximity to areas with wind or solar energy potential, accessibility to power transmission lines, and natural gas networks. The authors identified sites that could act as reservoirs for underground gas storage (of hydrogen or methane) (UGS), compressed air energy storage (CAES), underground pumped hydro energy storage (UPHES), and underground thermal energy storage (UTES); they concluded that, for the Portuguese geological context, the technologies with best application potential seem to be CAES and UGS linked to PtG.

Despite the envisaged underground storage potential yet to be explored, perspectives for PtG deployment in Portugal would be even more promising with the construction of the third planned connection with Spain in the natural gas network and of projected connections linking the Iberian Peninsula to France.

3.4. Research Needs

The current challenges regarding the technologies involved in PtG processes were extensively addressed in recent publications (e.g., [2,9,11,44]), and for that reason were out of the scope of this work, although their study has been the main focus of previous authors (e.g., [45,46]) and future research activities [19]. Instead, the present work aimed to provide a picture of the recent evolution of the Portuguese energy sector, highlighting the tremendous endeavor and commitment of the country for large-scale renewable energy deployment and to raise awareness about what seems to be promising conditions for PtG deployment. Nevertheless, future research studies to forecast surplus power in different energy case scenarios and the identification and characterization of the most suitable CO_2 point sources, besides techno-economic-environmental assessments, will be crucial to find profitable business models and integrated value chains for PtG deployment in Portugal. Among them, process chains should be looked at, leading to opportunities for O_2 (by-product of H_2O electrolysis) valorization, recycling of H_2O from CO_2 methanation, and energy integration to tackle current barriers for commercialization of PtG systems.

4. Conclusions

The present analysis of the Portuguese energy sector highlights the country's intense dependence on fossil fuels to afford its energetic needs, although, despite this, it was the fourth EU-18 member with the highest incorporation of renewables in power production in 2015 (i.e., 44.6%), a value that reached 57% in 2016 [36]. So far, the country's options to manage the energy surpluses generated by electricity from renewable sources relies on pumped hydro storage plants or power exportation to Spain. Hence, decentralized power-to-methane applications can be of strategic relevance for the country, since power production from natural gas will increase following the decommissioning of the Sines and Pego coal power plants by 2021. Storing surplus renewable electricity as methane would also allow the diversification of natural gas provision, minimizing the dependence and risk of shortage supply from foreign countries, as it is advised by the Portuguese Directorate-General for Energy and

Geology [47]. Additionally, a significant increase of natural gas consumption in the 2005–2015 decade (ca. 19%) by several and important industry sectors was also shown.

Portugal has important geographic advantages in favor of PtM demonstration projects, such as a well-developed natural gas network near wind parks and CO_2 sources, as well as a promising underground storage potential yet to be explored. For such a purpose, the engagement of all stakeholders (namely, academics, governmental bodies, technology and energy providers, major CO_2 polluting companies, and natural gas consumers) will be crucial for establishing national and/or regional research and development roadmaps, where the barriers (e.g., technical, legal, and regulatory), challenges, and opportunities for fast PtG deployment should be identified for coordinated actions.

Author Contributions: Conceptualization, C.V.M.; Writing—original draft preparation, C.V.M.; writing—review and editing, A.M. and L.M.M.; supervision, A.M. and L.M.M; Funding acquisition, A.M. and L.M.M.

Funding: C.V. Miguel is grateful to the Portuguese Foundation for Science and Technology (FCT) for his PhD scholarship (SFRH/BD/110580/2015), financed by national funds of the Ministry of Science, Technology and Higher Education and the European Social Fund (ESF) through the Human Capital Operational Programme (POCH). The authors acknowledge financial support from projects: (i) POCI-01-0145-FEDER-006939 (Laboratory for Process Engineering, Environment, Biotechnology and Energy—UID/EQU/00511/2013) funded by the European Regional Development Fund (ERDF), through COMPETE2020—Programa Operacional Competitividade e Internacionalização (POCI) and by national funds, through FCT—Fundação para a Ciência e a Tecnologia; (ii) NORTE-01-0145-FEDER-000005—LEPABE-2-ECO-INNOVATION, supported by North Portugal Regional Operational Programme (NORTE 2020), under the Portugal 2020 Partnership Agreement, through the European Regional Development Fund (ERDF); (iii) POCI-01-0145-FEDER-030277—funded by the ERDF funds through COMPETE2020—Programa Operacional Competitividade e Internacionalização (POCI) and by national funds (PIDDAC) through FCT/MCTES.

Conflicts of Interest: The authors declare no conflict of interest.

References

1. European Commission. Renewable Energy Road Map-Renewable Energies in the 21st Century: Building a More Sustainable Future. Available online: http://eur-lex.europa.eu/legal-content/EN/TXT/PDF/?uri=CELEX:52006DC0848&from=EN (accessed on 11 October 2018).
2. Götz, M.; Lefebvre, J.; Mörs, F.; McDaniel Koch, A.; Graf, F.; Bajohr, S.; Reimert, R.; Kolb, T. Renewable Power-to-Gas: A technological and economic review. *Renew. Energy* **2016**, *85*, 1371–1390.
3. Direção Geral de Energia e Geologia. Renováveis-Estatísticas Rápidas. Available online: http://www.dgeg.gov.pt (accessed on 7 November 2018).
4. Gomes, J.F.P. Reflections on the use of renewable power sources and nuclear energy in Portugal. *Int. J. Environ. Stud.* **2008**, *65*, 755–767. [CrossRef]
5. Mateus, C.B.; Estanqueiro, A. Regulation of the wind power production: Contribution of the electric vehicles and other energy storage systems. In Proceedings of the 11th International Workshop on Large-Scale Integration of Wind Power into Power Systems As Well As on Transmission Networks for Offshore Power Plants, Lisboa, Portugal, 13–15 November 2012.
6. Heymann, F.; Bessa, R. Power-to-Gas potential assessment of Portugal under special consideration of LCOE. In Proceedings of the 2015 IEEE PowerTech Eindhoven, Eindhoven, The Netherlands, 29 June–2 July 2015.
7. Zakeri, B.; Syri, S. Electrical energy storage systems: A comparative life cycle cost analysis. *Renew. Sustain. Energy Rev.* **2015**, *42*, 569–596. [CrossRef]
8. Blanco, H.; Faaij, A. A review at the role of storage in energy systems with a focus on Power to Gas and long-term storage. *Renew. Sustain. Energy Rev.* **2018**, *81*, 1049–1086. [CrossRef]
9. Maroufmashat, A.; Fowler, M. Transition of Future Energy System Infrastructure; through Power-to-Gas Pathways. *Energies* **2017**, *10*, 1089. [CrossRef]
10. Bailera, M.; Lisbona, P.; Romeo, L.M.; Espatolero, S. Power to Gas projects review: Lab, pilot and demo plants for storing renewable energy and CO_2. *Renew. Sustain. Energy Rev.* **2017**, *69*, 292–312. [CrossRef]
11. Eveloy, V.; Gebreegziabher, T. A Review of Projected Power-to-Gas Deployment Scenarios. *Energies* **2018**, *11*, 1824. [CrossRef]

12. Specht, M.; Brellochs, J.; Frick, V.; Stürmer, B.; Zuberbühler, U. The Power to Gas Process: Storage of Renewable Energy in the Natural Gas Grid via Fixed Bed Methanation of CO_2/H_2. In *Synthetic Natural Gas from Coal, Dry Biomass, and Power-to-Gas Applications*; Schildhauer, T.J., Biollaz, S.M.A., Eds.; John Wiley and Sons, Inc.: Hoboken, NJ, USA, 2016.
13. Quarton, C.J.; Samsatli, S. Power-to-gas for injection into the gas grid: What can we learn from real-life projects, economic assessments and systems modelling? *Renew. Sustain. Energy Rev.* **2018**, *98*, 302–316. [CrossRef]
14. Kitasei, S. *Powering the Low-Carbon Economy: The Once and Future Roles of Renewable Energy and Natural Gas*; Worldwatch Institute: Washington, DC, USA, 2010.
15. Walspurger, S.; Haije, W.G.; Louis, B. CO_2 reduction to substitute natural gas: Toward a global low carbon energy system. *Israel J. Chem* **2014**, *54*, 1432–1442. [CrossRef]
16. Mellquist, N.; Fulton, M. *Natural Gas and Renewables: A Secure Low Carbon Future Energy Plan for the United States*; Deutsche Bank Group: Frankfurt am Main, Germany, 2010.
17. German Energy Agency. Power to gas System Solution. Opportunities, Challenges and Parameters on the Way to Marketability. Available online: http://europeanpowertogas.com/wp-content/uploads/2018/05/wcX8e2Jv.pdf (accessed on 7 november 2018).
18. International Renewable Energy Agency. Renewable Energy Prospects: Germany. November 2015. Available online: http://www.irena.org/DocumentDownloads/Publications/IRENA_REmap_Germany_report_2015.pdf (accessed on 7 November 2018).
19. Power2Methane Project. Available online: http://power2methane.fe.up.pt (accessed on 11 November 2018).
20. Carneiro, J.F.; Matos, C.R.; van Gessel, S. Opportunities for large-scale energy storage in geological formations in mainland Portugal. *Renew. Sustain. Energy Rev.* **2019**, *99*, 201–211. [CrossRef]
21. Bento, N.; Fontes, M. The capacity for adopting energy innovations in Portugal: Historical evidence and perspectives for the future. *Technol. Forecast. Soc. Chang.* **2016**, *113*, 308–318. [CrossRef]
22. Eurostat, Environment and Energy Database. Available online: http://ec.europa.eu/eurostat/data/database, (accessed on 11 October 2018).
23. International Energy Agency. *Energy Policies of IEA Countries: Portugal 2009 Review*; OECD/IEA: Paris, France, 2009.
24. International Energy Agency. *Energy Policies of IEA Countries: Portugal 2016 Review*; OECD/IEA: Paris, France, 2016.
25. Amador, J. Energy Production and Consumption in Portugal: Stylized Facts. Available online: https://www.bportugal.pt/sites/default/files/anexos/papers/ab201007_e.pdf (accessed on 7 November 2018).
26. Direção Geral de Energia e Geologia. Fatura Energética Portuguesa 2016. Available online: http://www.dgeg.gov.pt (accessed on 07 November 2018).
27. Eurostat. Energy Balances. Available online: http://ec.europa.eu/eurostat/web/energy/data/energy-balances (accessed on 11 October 2018).
28. Direção Geral de Energia e Geologia. Relatório Sobre Avaliação dos Riscos que afetam o aprovisionamento de Gás Natural em Portugal-Período 2017–2025. Available online: http://www.dgeg.gov.pt/wwwbase/wwwinclude/ficheiro.aspx?access=1&id=15863 (accessed on 07 November 2018).
29. Entidade Nacional para o Mercado de Combústiveis. Mercado de Biocombustíveis em Portugal: A atuação da ENMC Desde Abril de 2015. Available online: http://www.enmc.pt/static-img/2016-03/2016-03-16101508_f7664ca7-3a1a-4b25-9f46-2056eef44c33$$72f445d4-8e31-416a-bd01-d7b980134d0f$$ff280618-fbff-40a0-b20c-a610d6977312$$File$$pt$$1.pdf (accessed on 11 October 2018).
30. Direção Geral de Energia e Geologia. Renováveis-Estatísticas Rápidas, Outubro 2014. Available online: http://www.dgeg.gov.pt (accessed on 7 November 2018).
31. Carvalho, J.M.; Coelho, L.; Nunes, J.C.; do Rosário Carvalho, M.; Garcia, J.; Cerdeira, R. Portugal Country Update 2015. In Proceedings of the World Geothermal Congress 2015, Melbourne, Australia, 19–25 April 2015.
32. Ferreira, S.; Monteiro, E.; Brito, P.; Vilarinho, C. Biomass resources in Portugal: Current status and prospects. *Renew. Sustain. Energy Rev.* **2017**, *78*, 1221–1235. [CrossRef]
33. Carvalho, D.; Wemans, J.; Lima, J.; Malico, I. Photovoltaic energy mini-generation: Future perspectives for Portugal. *Energy Policy* **2011**, *39*, 5465–5473. [CrossRef]

34. Teotónio, C.; Fortes, P.; Roebeling, P.; Rodriguez, M.; Robaina-Alves, M. Assessing the impacts of climate change on hydropower generation and the power sector in Portugal: A partial equilibrium approach. *Renew. Sustain. Energy Rev.* **2017**, *74*, 788–799.
35. Rehman, S.; Al-Hadhrami, L.M.; Alam, M.M. Pumped hydro energy storage system: A technological review. *Renew. Sustain. Energy Rev.* **2015**, *44*, 586–598. [CrossRef]
36. Redes Eletricas Nacionais, Technical Data 2016. Available online: http://www.centrodeinformacao.ren. pt/PT/InformacaoTecnica/DadosTecnicos/REN%20Dados%20T%C3%A9cnicos%202016.pdf (accessed on 11 October 2018).
37. Silva, B.; Marques, M.; Matos, J.C.; Rodrigues, A.; Pereira, R.; Bastos, A.; Cabral, P.; Saraiva, F. Wind power in Portugal: From the potential to the integration in the electric grid. Prooceedings of the European Wind Energy Conference and Exhibition, EWEC 2010, Warsaw, Poland, 20–23 April 2010; pp. 4508–4515.
38. Marques, J.A.R. Hidroelectricidade e Barragens Reversíveis: Panorama Actual. Master's Thesis, Universidade do Porto, Porto, Portugal, 2015.
39. Peña, I.; Azevedo, I.; Marcelino Ferreira, L.A.F. Lessons from wind policy in Portugal. *Energy Policy* **2017**, *103*, 193–202. [CrossRef]
40. Pereira, T.C.; Seabra, T.; Pina, A.; Canaveira, P.; Amaro, A.; Borges, M.; Silva, R. *Portuguese National Inventory Report on Greenhouse Gases, 1990–2015 Submmitted under the United Nations Framework Convention on Climate Change and the Kyoto Protocol*; Portuguese Environmental Agency: Amadora, Portugal, 2017.
41. Carneiro, J.F.; Boavida, D.; Silva, R. First assessment of sources and sinks for carbon capture and geological storage in Portugal. *Int. J. Greenh. Gas Control* **2011**, *5*, 538–548. [CrossRef]
42. KBB Underground Technologies GmbH. *Underground Gas Storage Carriço/Portugal*; KBB Underground Technologies GmbH: Hannover, Germany, 2015.
43. Nunes, P.V.C. Potencial de Armazenamento Subterrâneo em Cavidades Salinas de Gás Natural em Portugal. Master's Thesis, Instituto Superior Técnico, Lisboa, Portugal, 2010.
44. Buttler, A.; Spliethoff, H. Current status of water electrolysis for energy storage, grid balancing and sector coupling via power-to-gas and power-to-liquids: A review. *Renew. Sustain. Energy Rev.* **2018**, *82*, 2440–2454. [CrossRef]
45. Miguel, C.V.; Soria, M.A.; Mendes, A.; Madeira, L.M. A sorptive reactor for CO_2 capture and conversion to renewable methane. *Chem. Eng. J.* **2017**, *322*, 590–602. [CrossRef]
46. Miguel, C.V.; Mendes, A.; Madeira, L.M. Intrinsic kinetics of CO_2 methanation over an industrial nickel-based catalyst. *J. CO_2 Util.* **2018**, *25*, 128–136. [CrossRef]
47. Direção Geral de Energia e Geologia. Relatório de Monitorização da Segurança de Abastecimento do Sistema Elétrico Nacional 2013–2030. Available online: http://www.erse.pt/pt/consultaspublicas/consultas/ Documents/49_1/RMSA-E%202012.pdf (accessed on 7 November 2018).

Article

Comparison of Technologies for CO_2 Capture from Cement Production—Part 1: Technical Evaluation

Mari Voldsund [1,*], Stefania Osk Gardarsdottir [1], Edoardo De Lena [2], José-Francisco Pérez-Calvo [3], Armin Jamali [4], David Berstad [1], Chao Fu [1], Matteo Romano [2], Simon Roussanaly [1], Rahul Anantharaman [1], Helmut Hoppe [4], Daniel Sutter [3], Marco Mazzotti [3], Matteo Gazzani [5], Giovanni Cinti [6] and Kristin Jordal [1]

[1] SINTEF Energy Research, 7465 Trondheim, Norway; stefania.gardarsdottir@sintef.no (S.O.G.); david.berstad@sintef.no (D.B.); chao.fu@sintef.no (C.F.); simon.roussanaly@sintef.no (S.R.); rahul.anantharaman@sintef.no (R.A.); kristin.jordal@sintef.no (K.J.)
[2] Politecnico di Milano, Department of Energy, 20156 Milan, Italy; edoardo.delena@polimi.it (E.D.L.); matteo.romano@polimi.it (M.R.)
[3] ETH Zurich, Institute of Process Engineering, 8092 Zurich, Switzerland; francisco.perezcalvo@ipe.mavt.ethz.ch (J.-F.P.-C.); sutter@ipe.mavt.ethz.ch (D.S.); marco.mazzotti@ipe.mavt.ethz.ch (M.M.)
[4] VDZ gGmbH, 40476 Düsseldorf, Germany; armin.jamali@vdz-online.de (A.J.); helmut.hoppe@vdz-online.de (H.H.)
[5] Utrecht University, Copernicus Institute of Sustainable Development, 3584 CB Utrecht, The Netherlands; m.gazzani@uu.nl
[6] Italcementi Heidelberg Group, 24126 Bergamo, Italy; g.cinti@italcementi.it
[*] Correspondence: mari.voldsund@sintef.no; Tel.: +47-907-48-259

Received: 6 December 2018; Accepted: 4 February 2019; Published: 12 February 2019

Abstract: A technical evaluation of CO_2 capture technologies when retrofitted to a cement plant is performed. The investigated technologies are the oxyfuel process, the chilled ammonia process, membrane-assisted CO_2 liquefaction, and the calcium looping process with tail-end and integrated configurations. For comparison, absorption with monoethanolamine (MEA) is used as reference technology. The focus of the evaluation is on emission abatement, energy performance, and retrofitability. All the investigated technologies perform better than the reference both in terms of emission abatement and energy consumption. The equivalent CO_2 avoided are 73–90%, while it is 64% for MEA, considering the average EU-28 electricity mix. The specific primary energy consumption for CO_2 avoided is 1.63–4.07 MJ/kg CO_2, compared to 7.08 MJ/kg CO_2 for MEA. The calcium looping technologies have the highest emission abatement potential, while the oxyfuel process has the best energy performance. When it comes to retrofitability, the post-combustion technologies show significant advantages compared to the oxyfuel and to the integrated calcium looping technologies. Furthermore, the performance of the individual technologies shows strong dependencies on site-specific and plant-specific factors. Therefore, rather than identifying one single best technology, it is emphasized that CO_2 capture in the cement industry should be performed with a portfolio of capture technologies, where the preferred choice for each specific plant depends on local factors.

Keywords: CO_2 capture; cement production with CO_2 capture; CO_2 capture in industry; CO_2 capture retrofitability; oxyfuel; chilled ammonia; membrane-assisted CO_2 liquefaction; calcium looping

---

1. Introduction

Production of cement is estimated to account for around 7% of global CO_2 emissions (2018) [1]. The cement industry can reduce its specific CO_2 emissions through a variety of different techniques, such as increased energy efficiency, utilization of alternative fuels, application of alternative raw materials, and reduction of the clinker to cement ratio. However, these techniques have already been exploited to a significant extent, and they can only partly reduce the emissions [2]. Around two-thirds of the CO_2 emissions from the cement industry are process related, originating from the calcination of limestone where $CaCO_3$ is converted to CaO and CO_2, while one-third of the emissions come from combustion of fuels in the cement plant's calciner and rotary kiln. A measure such as fuel switch can therefore only remove one-third of the CO_2 emissions. CO_2 capture and storage (CCS) can significantly reduce both the process related and fuel related emissions. It is identified as the single measure that has the largest potential for further overall emission reductions in the cement industry [1,2].

Cement kilns usually have a lifetime of 30–50 years [3]. Although a few kilns might have to be rebuilt to meet EU requirements on pollutant emissions and technical standards [4], it is not likely that many kilns are to be built in Europe in a foreseeable future, as the cement production is anticipated to be approximately constant in the next decades [1]. Therefore, in order to abate CO_2 emissions from European cement production, it is important that CO_2 capture technologies can be retrofitted to existing cement plants.

The most mature CO_2 capture technology at present is chemical absorption with amine. However, this technology may not necessarily be the best option when alternative technologies have matured. In literature, several CO_2 capture technologies have been evaluated for use in the cement industry. Most of the studies have focused on amine-based CO_2 capture [2,5–9], followed by fewer studies on calcium looping [6,10,11], oxyfuel [2,8], and membrane-based technologies [6,7,12]. Comprehensive work has been performed in each of the studies, but technical and economic performance cannot be directly compared between studies since they are based on different assumptions. A consistent evaluation of more than two types of CO_2 capture technologies for the cement industry is not available.

In most of the assessments of CO_2 capture technologies in the cement industry, retrofit to an existing cement plant is assumed. However, there are several practical aspects linked to retrofitability that are not necessarily reflected in energy performance and cost. For instance, aspects such as space requirement or added load to the local power grid could potentially determine whether a technology can be implemented or not. Liang and Li [5] defined a list of criteria for assessing the potential of a cement plant to be retrofitted with CO_2 capture technologies. However, there is to the authors' knowledge no such work done on assessing retrofitability of capture technologies to cement plants.

This paper presents a consistent technical evaluation of CO_2 capture technologies for retrofit in the cement industry. The evaluation focusses on emission abatement, energy performance, and retrofitability. The following technologies are investigated:

- oxyfuel process
- chilled ammonia process
- membrane-assisted CO_2 liquefaction
- calcium looping (tail-end and integrated configuration)

These technologies comprise a set of fundamentally different technologies that appear promising for application in the cement sector. Absorption with monoethanloamine (MEA) is used as reference technology due to its benchmark status in the literature.

This work has been carried out as a part of the Horizon 2020 project "CO_2 capture from cement production" (CEMCAP) [13], which has the overall objective to prepare the ground for large-scale implementation of CO_2 capture in the European cement industry. An essential element in responding to this objective has been to perform a comprehensive techno-economic comparative assessment of CO_2 capture, which can be used as a decision basis for future evaluations of CO_2 capture implementation at cement plants. An extraction of this work is presented as a paper series, where the technical evaluation presented in the current paper forms Part 1, and an economic analysis forms Part 2.

2. Reference Cement Plant and CO_2 Capture Technologies

2.1. Reference Cement Plant

The reference cement plant is a Best Available Technique (BAT) plant defined by the so-called European Bref document [14] and adopted by the European Cement Research Academy (ECRA) and CEMCAP [2,9,15]. It is based on a dry kiln process, consisting of a five-stage cyclone preheater, calciner with tertiary air duct, rotary kiln, and grate cooler. The most important characteristics of the plant are summarized in Table 1. It has a capacity of 3.0 kt clinker per day, which corresponds to ca. 1.0 Mt clinker per year, or 1.4 Mt cement per year, with a run time of >330 days per year.

Table 1. Characteristics of the reference cement plant.

Parameter	Value
Clinker production	3000 t/day
Clinker/cement factor	0.737
Raw meal/clinker factor	1.6
Specific CO_2 emissions	850 kg/tclk
Specific electric power consumption	97 kWh/tcement

The clinker burning line of the plant is shown in Figure 1. The raw material, which consists of 77 wt% $CaCO_3$, 14 wt% SiO_2, and small amounts of Al_2O_3, Fe_2O_3, $MgCO_3$, and water, is first ground in the raw mill to form raw meal, where it is also dried by hot flue gas from the preheater. The flue gas and the raw meal are subsequently separated in a dust filter, and the raw meal is sent to the preheater while the flue gas is sent to the stack. In the preheater the meal is heated by hot flue gas coming from the calciner and the rotary kiln. The meal and the hot gases are first mixed for heat transfer and then separated again in cyclones arranged above one another. The preheated raw meal enters the calciner, where the major part of the calcination ($CaCO_3$ -> CaO + CO_2) is performed. Around 60% of the plant's total fuel input is consumed here to achieve a suitable temperature (~860 °C) and drive the endothermic reaction. After the calciner, the calcined (>95%) raw meal enters the rotary kiln, where the still unreacted part of the limestone is calcined after a few meters and the formation of clinker takes place. The rotary kiln is heated by the main burner where the remaining 40% of the plant's fuel input is consumed. The solid material reaches 1450 °C, and the temperature of the gas phase can reach 2000 °C. The hot clinker is discharged from the kiln to a clinker cooler, where the clinker is cooled with ambient air. Some of the resulting hot air is used as combustion air in the main burner (secondary air) and in the calciner (tertiary air).

Emissions of NO_x are controlled by selective non-catalytic reduction (SNCR) to not exceed the permitted limit of 500 mg/Nm^3 at 10% O_2 [16]. In practice this means that ammonia solution is injected in the kiln system to reduce NO_x to N_2. No system is installed for SO_x emission control, because the SO_x emissions are already below the permitted limit. The CEMCAP reference plant is identical to the ECRA reference plant, with the exception that SNCR is assumed to be implemented in the CEMCAP reference plant and not in the ECRA reference plant.

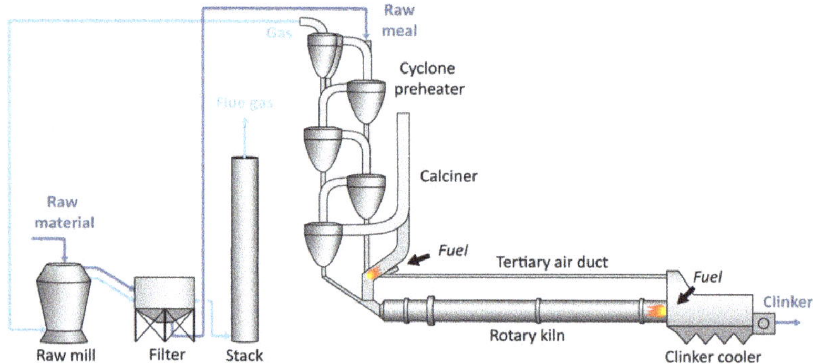

Figure 1. The clinker burning line of the reference cement plant.

The composition and flow rate of the flue gas depends on the amount of air leaking into the system. The nominal air leak increases with time and operation of the plant, while it is reinstated via the yearly plant maintenance. The air leak in the raw mill is predominant, so variation over the year is only considered after the mill. Flue gas conditions at the stack during the first and second part of the year are given in Table 2.

Table 2. Flue gas conditions at stack. Nm^3 refers to normal cubic meters (volume at 0 °C and 1 atm).

	First $\frac{1}{2}$ Year	Second $\frac{1}{2}$ Year
Air leak in mill	Low	Medium
False air at preheater outlet (vol%)	5.5	5.5
False air at stack (vol%)	30	43
Total flow rate (kg/h)	318,192	388,098
Temperature (°C)	130	110
Mole fraction (wet basis)		
CO_2	0.22	0.18
N_2	0.60	0.63
O_2	0.07	0.10
H_2O	0.11	0.09
NO_x (dry basis) (mg/Nm^3)	591 [1]	455 [1]
SO_x (dry basis) (mg/Nm^3)	236 [2]	182 [2]
Dust (dry basis, at 10% O_2) (mg/Nm^3)	10	10

[1] Corresponds to 500 mg/Nm^3 at 10% O_2. [2] Corresponds to 200 mg/Nm^3 at 10% O_2.

Waste heat can be recovered from the cooler exhaust air to produce steam (Figure 2). The dust in the exhaust air is assumed to be removed with a ceramic filter prior to the heat recovery steam generator, which is assumed to have a minimum approach temperature of 80 °C.

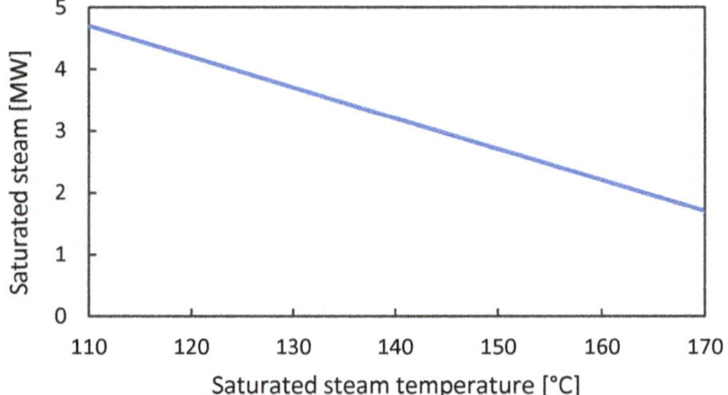

Figure 2. Maximum amount of saturated steam that can be generated by heat recovery from the cooler exhaust air as function of the saturated steam temperature, considering a minimum approach temperature of 80 °C.

2.2. MEA Absorption

The reference technology MEA absorption is a post-combustion technology, where CO_2 is absorbed from the flue gas with MEA solvent (Figure 3). To limit solvent degradation, the content of NO_x and SO_x in the flue gas must be reduced beyond the permitted emission limits before the flue gas comes in contact with the solvent. The already existing SNCR system is assumed to be utilized to remove additional NO_x before the flue gas leaves the kiln. This flue gas is then cooled in a direct contact cooler (DCC) where water also is removed, and SO_x is removed by scrubbing with NaOH. The cooled flue gas can then be sent to the absorber column where aqueous MEA solution (30 wt %) absorbs CO_2 from the gas. Evaporated MEA is recovered from the purified flue gas in a water wash section at the top of the absorber. The CO_2 rich MEA solvent is regenerated in a desorber column, and the resulting high-purity CO_2 is compressed to reach the transport specifications.

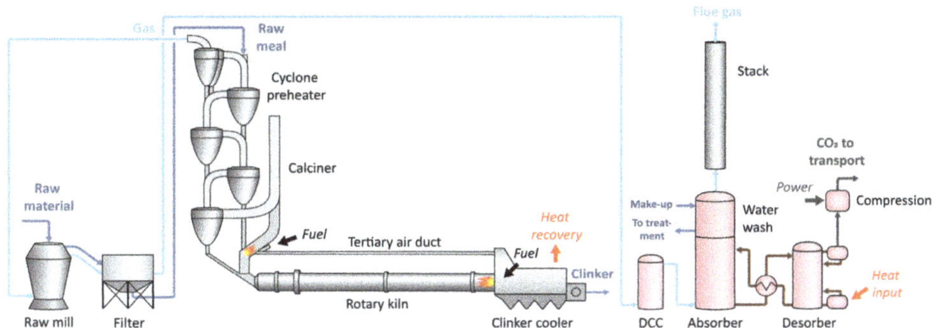

Figure 3. Reference clinker burning line with MEA absorption.

The process requires a considerable amount of heat for solvent regeneration, and power is required for fans and pumps in the absorption process and for compression of the captured CO_2. Waste heat can be used to cover some of the heat demand. For the reference cement plant, the available waste heat can cover 4% of the total heat demand.

2.3. Oxyfuel Process

In the oxyfuel process (Figure 4), combustion is performed with an oxidizer consisting mainly of oxygen mixed with recycled CO_2, to produce a CO_2 rich flue gas which allows a relatively easy purification with a CO_2 purification unit (CPU). As opposed to the MEA technology, the cement kiln process itself must be modified. The gas atmosphere in the clinker cooler, the rotary kiln, the calciner and the preheater is changed, and some of the flue gas is recycled. Air that is heated by hot gases from the preheater and the clinker cooler is sent to the raw mill to dry the raw material, instead of the flue gas.

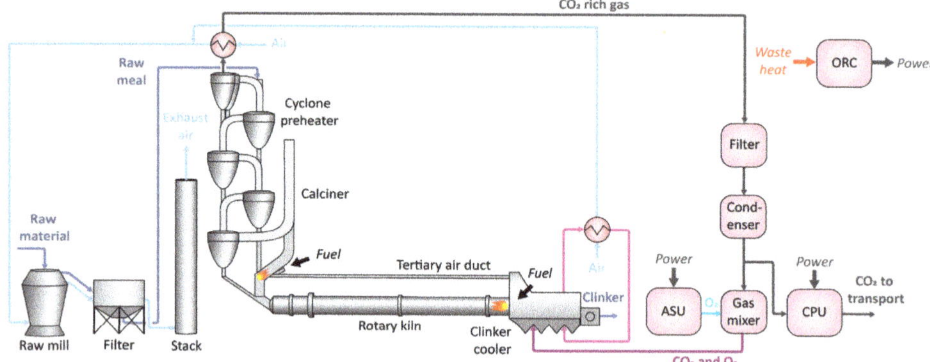

Figure 4. Reference clinker burning line with oxyfuel CO_2 capture.

Additional power is required for the oxyfuel process compared to a plant without capture, mainly by an air separation unit (ASU) providing oxygen and the CPU. Some of this power demand can be covered by a waste heat recovery system. In this analysis, an organic Rankine cycle (ORC) is assumed to be installed.

2.4. Chilled Ammonia Process

The chilled ammonia process (CAP) is a post-combustion technology based on absorption, where CO_2 is removed from the flue gas using chilled ammonia as solvent (Figure 5). The flue gas is first conditioned in a DCC where it is cooled and SO_x is removed by scrubbing with ammonia, before it is sent to an absorption column where CO_2 is removed by an ammonia solution. The temperature in the absorber is controlled by a solvent pump-around that is chilled down to temperatures around 12–13 °C. Ammonia is recovered from the flue gas in a water wash section at the top of the absorber, before purified flue gas is released to the atmosphere. Ammonia is desorbed from the wash water in a desorption column and recycled into the process. CO_2 rich ammonia solvent is regenerated in a CO_2 desorber that is operated at around 25 bar. The resulting high-purity CO_2 stream must be further pressurized to meet the transport specifications.

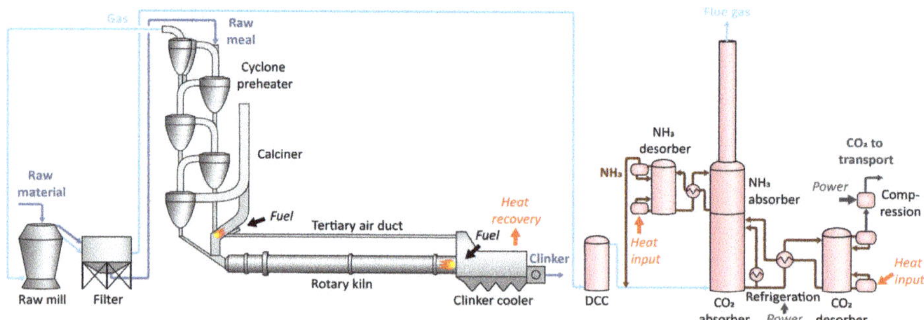

Figure 5. Reference clinker burning line with CAP CO_2 capture.

In this process heat is required for solvent regeneration and for the ammonia recovery system, and power is required for chilling, pumping, and compression. Waste heat can be utilized to cover a part of the heat demand. This amounts to 7–8% of the total heat demand in the case of the reference cement plant.

2.5. Membrane-Assisted CO_2 Liquefaction

The membrane-assisted CO_2 liquefaction (MAL) technology comprises a combination of polymeric membranes and a CO_2 liquefaction process (Figure 6). The polymeric membranes are utilized for bulk separation resulting in CO_2 a product with moderate purity. This product is further treated in the CO_2 liquefaction process, where CO_2 is liquefied to form high purity CO_2, while the partially decarbonized tail gas is recycled to the membrane feed gas.

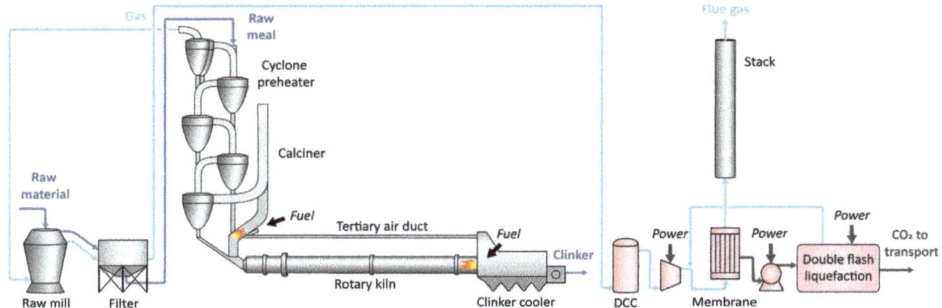

Figure 6. Reference clinker burning line with membrane-assisted CO_2 liquefaction.

The flue gas entering the process is first cooled and water is removed in a DCC, before it is compressed and sent to the membrane module. The pressure difference and pressure ratio over the membrane module is generated both by the flue gas compression on the feed side and vacuum pumps on the permeate side of the membrane. The chemical stability of polymeric membranes depends on the type of polymer, and the tolerance of such membranes to SO_x and NO_x is often highlighted as uncertain [17]. However, some membrane producers report on high tolerances for their membranes [18]. In this work, the same strategy for SO_x and NO_x removal prior to the capture system as for MEA absorption is assumed.

The technology is a post-combustion technology with no additional integration or feedback to the cement plant. Only electric power is required as input to the process.

2.6. Calcium Looping—Tail-End Configuration

The calcium looping (CaL) technology is based on the reversible carbonation reaction $(CaO + CO_2 \rightleftharpoons CaCO_3)$, which is exploited to separate carbon dioxide from flue gas. The technology can be applied to a cement plant in a tail-end configuration (Figure 7) or it can be integrated with the calcination process in the cement kiln. In the tail-end configuration the flue gas from the cement kiln is sent to a carbonator where CO_2 is removed by reaction with CaO-based sorbent. The sorbent is regenerated in a calciner where coal is burnt under oxyfuel conditions to reach the calcination temperature of around 920 °C. The captured CO_2 requires some additional purification in a CPU. CaO-rich purge from the system is sent to the cement kiln and used as constituents of the raw meal.

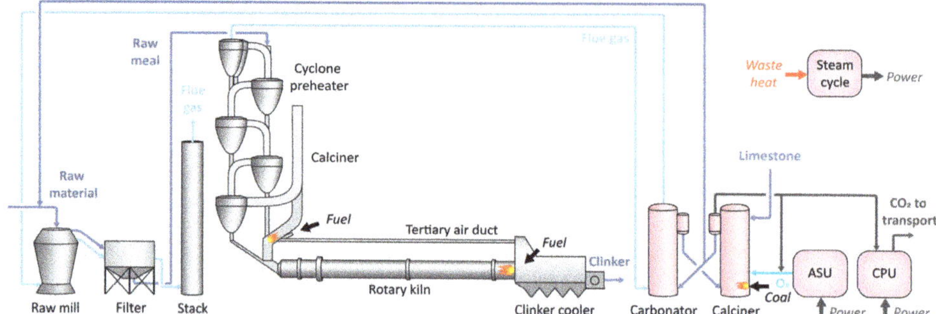

Figure 7. Reference clinker burning line with calcium looping CO_2 capture—tail-end configuration.

The tail-end CaL process requires supply of limestone and coal. Power is required for an ASU providing oxygen, fans, and the CPU. A steam cycle recovers high temperature waste heat and produces power. Depending on the case this results in net import or net export of power.

2.7. Calcium Looping—Integrated Configuration

In the integrated calcium looping configuration, the CO_2 capture calciner is combined with the calciner of the cement kiln for a more energy efficient process (Figure 8). Because of the small size of the particles required in the calciner of the cement kiln, which make them hardly fluidizable, the carbonation and calcination reactions must take place in entrained flow reactors instead of circulating fluidized bed reactors that are normally used in CaL processes.

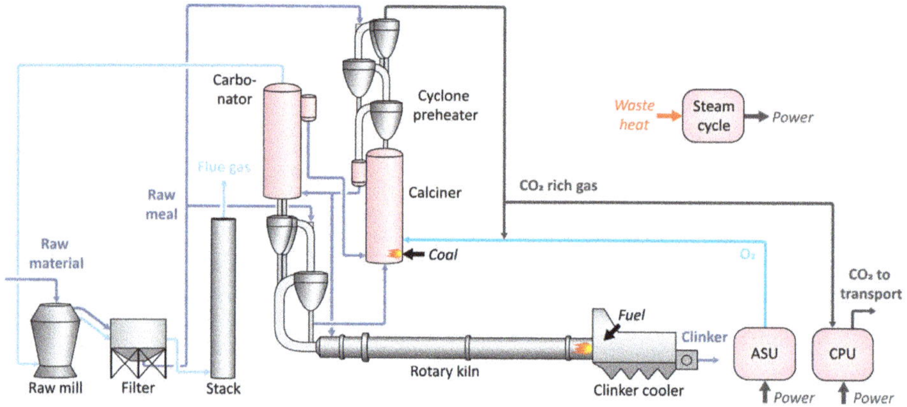

Figure 8. Reference clinker burning line with calcium looping CO_2 capture—integrated configuration.

As for the tail-end configuration, additional fuel is required, and the calciner must be operated under oxyfuel conditions. Power is required for an ASU, a CPU and fans, but power is also generated by a steam cycle utilizing waste heat in the process.

3. Methodology

Process simulations of the reference plant and each of the technologies are established as a basis for the analysis. Common assumptions for the process simulations such as specifications for different types of process units are used as defined in the CEMCAP framework [15], and the simulations are aligned with experimental research performed within the project [19–26].

The oxyfuel and calcium looping technologies are closely integrated with the kiln, while the other technologies are only connected to the kiln by the flue gas entering the system and heat integration if heat is required. Due to the integration with the kiln, the two partners simulating these processes, VDZ, and Politecnico di Milano (PoliMi), have established their own simulations of the reference kiln. The other technologies are simulated using flue gas from the VDZ simulation of the reference kiln as feed to the process.

The captured CO_2, direct and indirect CO_2 emissions, and primary energy consumption of the reference cement plant with and without CO_2 capture are quantified based on the process simulations and used to calculate quantitative key performance indicators (KPIs) on emission abatement and energy performance. It should be noted that since simulations of MEA, oxyfuel, CAP, and MAL are based on the VDZ simulation of the reference kiln, values from this simulation are used as reference in the KPI calculations, while for the calcium looping technologies, values from the PoliMi simulation of the reference kiln are used as reference.

Finally, an assessment of retrofitability of all capture technologies is performed with respect to qualitative KPIs.

3.1. Common Design Specifications

All the capture processes are designed for minimum 90% CO_2 avoided from flue gas, AC_{fg}, and pipeline transport, which requires >95% CO_2, <300 ppm_{wt} H_2O, and 110 bar. For MEA, CAP, and MAL, which are installed after the raw mill, the process is designed for medium air leak and assumed to be operated with low air leak half of the year and medium air leak the other half of the year. The oxyfuel and CaL technologies are implemented before the mill, and the effect of change in air leak over the year is this neglected for these technologies.

3.2. Utility Energy Consumption and CO_2 Emissions

Calculated emissions and energy consumption of utilities are based on assumptions presented in Table 3. For the electricity generation efficiency and CO_2 emissions, the average electricity mix in EU-28 in 2014 is used as base case, but alternative scenarios that also are investigated are summarized in Table 4. For MEA and CAP, which require steam, it is assumed that the available waste heat will be used to cover the demand as far as possible, and that the rest is covered by a natural gas (NG) fired boiler.

Table 3. Assumptions on energy consumption (lower heating value) and CO_2 emissions related to utilities.

Parameter	Value
Coal (kJ/kg)	27,150
Natural gas specific CO_2 emissions (kg/GJ)	56.1
Natural gas boiler efficiency (%)	90
Electricity generation efficiency (EU 2014), η_{el} (%)	45.9
Electricity generation specific CO_2 emissions (EU 2014), e_{el} (kg/MWh)	262
Electric heater efficiency (%)	95
ASU power demand [27] (kWh/t O_2)	226
ASU cooling demand [28] (kJ/kg O_2)	566
ASU dehydration heat demand [28] (kJ/kg O_2)	58.3
CO_2 dehydration with TEG heat demand (kJ/kg CO_2)	2.62
CO_2 dehydration with TEG power demand (kJ/kg CO_2)	0.045
Cooling water system power demand (kW/kW)	0.02

Table 4. Generation efficiency η_{el} and specific CO_2 emissions e_{el} of electric power in alternative scenarios for electric power generation.

Power Generation Case	η_{el} (%)	e_{el} (kg/MWh)
Pulverized coal—ultra-supercritical state-of-the-art	44.2	770
Pulverized coal—sub-critical	35.0	973
Natural gas combined cycle	52.5	385
Renewables	100 [1]	0

[1] Physical energy content method as used by IEA and Eurostat.

3.3. Technical Key Performance Indicators

Key performance indicators for evaluation of emission abatement, energy performance, and retrofitability are defined.

3.3.1. Emission Abatement

The CO_2 capture ratio is a common KPI for CO_2 capture processes. It is defined as the CO_2 captured, $\dot{m}_{CO2,capt}$, divided by the CO_2 generated, $\dot{m}_{CO2,gen}$:

$$CCR = \frac{\dot{m}_{CO2,capt}}{\dot{m}_{CO2,gen}} \tag{1}$$

The CO_2 generated is both CO_2 generated in the cement kiln and CO_2 generated by fuel combustion internally in the capture process, but not CO_2 generated in the NG fired boilers or CO_2 generated indirectly by power consumption. The CO_2 captured is the total amount of CO_2 captured. There is no distinction between CO_2 originating from the cement process, and CO_2 originating from additional fuel combustion.

The CO_2 avoided from flue gas evaluates the direct CO_2 emission reduction from the flue gas of the cement kiln. It is defined as

$$AC_{fg} = \frac{e_{clk,fg,ref} - e_{clk,fg}}{e_{clk,fg,ref}} \tag{2}$$

where $e_{clk,fg,ref}$ denotes specific CO_2 emissions with the kiln flue gas in the reference plant, and $e_{clk,fg}$ denotes the specific CO_2 emissions with the flue gas of the kiln with capture. For technologies where a CPU is used, the CPU vent gas is accounted as a part of the flue gas. Emissions from steam generation (NG fired boilers) and indirect emissions related to power consumption/generation are not considered. This indicator differs from the CCR, because CO_2 captured from additional fuel combustion internally in the capture process is not counted as CO_2 avoided.

The equivalent CO_2 avoided evaluates the total equivalent CO_2 emissions avoided at the plant, taking the direct emissions related to the steam generation (NG fired boilers) in addition to the direct emissions with the flue gas, as well as the indirect CO_2 emissions associated to electric power consumption/generation, into account. It is defined as:

$$AC_{eq} = \frac{e_{clk,eq,ref} - e_{clk,eq}}{e_{clk,eq,ref}} \tag{3}$$

where $e_{clk,eq,ref}$ is specific equivalent emissions from the kiln without capture, and $e_{clk,eq}$ is the specific equivalent emission from the kiln with capture. Equivalent emissions are defined as the sum of direct e_{clk} and indirect $e_{el,clk}$ emissions

$$e_{clk,eq} = e_{clk} + e_{el,clk} \; [kg/t_{clk}] \tag{4}$$

Indirect emissions can be calculated using the equation

$$e_{el,clk} = e_{el} \cdot P_{el,clk} \ [kg/t_{clk}] \tag{5}$$

where $P_{el,clk}$ is the specific power consumption per unit mass of clinker produced, which is positive when power is consumed and negative when it is generated, and e_{el} is the CO_2 emissions associated with each unit of electric power consumed. This value depends on the electricity mix considered (see Table 4).

The equivalent CO_2 avoided takes all direct and indirect emissions into account. It gives the best indication on the overall reduction in CO_2 emissions of the cement plant when a certain capture technology is implemented and allows a fair comparison of different technologies.

3.3.2. Energy Performance

The specific primary energy consumption for CO_2 avoided, SPECCA, evaluates the primary energy used to avoid CO_2 emissions to the atmosphere. It is defined as the difference in equivalent primary energy consumption of the cement plant with and without CO_2 capture, divided by the difference in equivalent CO_2 emissions without and with capture

$$SPECCA = \frac{q_{clk,\ eq} - q_{clk,eq,ref}}{e_{clk,eq,ref} - e_{clk,eq}} \ [MJ/kg\ CO_2] \tag{6}$$

Equivalent specific primary energy consumption is the sum of direct (q_{clk}) and indirect ($q_{el,clk}$) specific primary energy consumption

$$q_{clk,\ eq} = q_{clk} + q_{el,clk} \ [MJ/t_{clk}] \tag{7}$$

The direct specific primary energy consumption is the amount of energy (lower heating value), supplied in the form of fuel (coal or natural gas), that is used per ton of clinker

$$q_{clk} = \frac{\dot{m}_{fuel} \cdot LHV_{fuel}}{\dot{m}_{clk}} \ [MJ/t_{clk}] \tag{8}$$

The indirect specific primary energy consumption is the amount of energy consumed by the generation of power required per ton of clinker. It can be calculated by

$$q_{el,clk} = \frac{P_{el,\ clk}}{\eta_{el}} \ [MJ/t_{clk}] \tag{9}$$

where η_{el} is the electricity generation efficiency, which depends on the electricity mix considered.

3.3.3. Retrofitability

To assess retrofitability, a qualitative assessment is performed. The following set of indicators is defined for the assessment:

- impact on the cement production process
- equipment and footprint
- utilities and services
- introduction of new chemicals/subsystems
- available operational experiences

These indicators evaluate different aspects that will affect the selection of capture technology for retrofit to an existing cement plant. Each of them is described in more detail below.

The evaluation is performed using the color-coding system presented in Table 5. For the color green, retrofit is evaluated as fairly straightforward; for yellow, retrofit is in most cases straightforward but some attention is needed; for orange more attention is needed, or important aspects are unknown so further assessment/research is needed; for red, a retrofit is evaluated not to be possible.

Table 5. Color coding for assessment of retrofitability.

Color	Explanation
	In most cases no or significantly less attention needed.
	Some attention needed for plant retrofit.
	Special attention needed for plant retrofit, or further assessment is needed.
	Retrofit is not possible.

3.4. Impact on the Cement Production Process

For a cement producer, the first priority is the production of high-quality clinker that can be ground and mixed with additives to produce cement. It is therefore very important that the installation of a CO_2 capture system does not influence the operability of the plant and the quality of the product. Some technologies imply modifications of the kiln system itself, and then the risk for effects on plant operability and product quality is increased. Furthermore, if fundamental modifications of the cement kiln system are required, long production stops during the installation of the technology may be required. This indicator evaluates the potential impact on clinker quality and plant operability, taking the phase with normal operation after installation as well as the installation phase into account.

3.5. Equipment and Footprint

The application of CO_2 capture technologies in cement plants requires the installation of additional equipment which demands a certain amount of space. The footprint of a typical cement plant with a 3000 t/d cement kiln is around 15 ha (150,000 m^2; a cement plant including the quarries can cover 100–150 ha). In general, if the new equipment can be installed anywhere at the plant, the space requirement would need some attention, but can in most cases be handled quite easily. However, in many cement plants the free space close to the kiln line is limited. If the new equipment must be installed near the kiln line, the footprint of the equipment could then be a limiting factor for individual capture technologies, and more attention to the characteristics of each site is required for assessing the applicability of the technology as a retrofit.

3.6. Utilities and Services

The application of certain CO_2 capture technologies requires additional utilities and services. This includes the need of additional electric power, steam (or NG for steam boilers), coal, or chemicals. Out of this, the most limiting factor is the electric power demand, since the availability is dependent on the local grid capacity. Infrastructure for import of natural gas and coal normally already exists at cement plants, and an increase in the import of these fuels can be handled easily up to a certain point, before the capacity of the infrastructure must be increased. In addition, some technologies require that the raw meal fed to the plant has certain properties. This indicator evaluates the technologies with respect to such aspects.

3.7. Introduction of New Chemicals/Subsystems

This indicator evaluates the attention needed concerning the introduction of new chemicals/subsystems at the plant. If new chemicals or subsystems are introduced, e.g., MEA, ammonia, oxygen, ASU, refrigeration system, ORC, or steam cycle, new procedures, and routines must be implemented to ensure safe operation. The utilization of new chemicals could also require additional permits. The relevant regulatory framework might vary from country to country. Constraints related to the handling of CO_2 at the plant are not included, since this will be the same for all technologies.

3.8. Available Operational Experiences

For the installation of CO_2 capture technologies at cement kilns, technology maturity is important to limit the associated risks. The assessment should take into account the available experiences with the application in the cement industry, but also the available experiences from other industrial sectors like the power industry.

4. Process Modeling and Key Process Data

In this section, the process modeling of the reference cement kiln and the capture technologies, as well as the most important process data are summarized. Stream data and PFDs of the complete systems, including CPUs and waste heat recovery systems, can be found in the Supplementary Materials.

4.1. Reference Cement Kiln

The core process of the reference cement kiln is simulated by VDZ with their in-house cement kiln process model [29,30], and by PoliMi with their in-house process simulation tool GS [31] (Table 6). The difference in specific heat input and specific CO_2 emissions between these two simulations is 2%. The effect of these differences will be considerably lower in the final quantitative KPIs, since values referring to the cement plant with and without CO_2 capture for the calculation of KPIs for each technology, will have been obtained using the same tool. Details on the simulations and a comparison of them are given by Campanari et al. [32].

Table 6. Key process data of reference cement kiln.

	VDZ Simulation	PoliMi Simulation
Clinker production (t/h)	120.6	117.7
Coal consumption (MW)	105.1	105.9
Electric power consumption [1] (MW)	15.9	15.5
CO_2 emissions at stack (t/h)	102.0	101.8

[1] Including raw meal and clinker grinding, solids handling, kiln drive, lightning, etc.

4.2. MEA Absorption

The MEA system is modelled as shown in Figure 9 using the process simulator Aspen HYSYS V9, with resulting key process data as summarized in Table 7. The acid gas property package is selected for modelling the part of the process including MEA solvent. The SRK property package is used for calculation of properties of the flue gas and CO_2 streams.

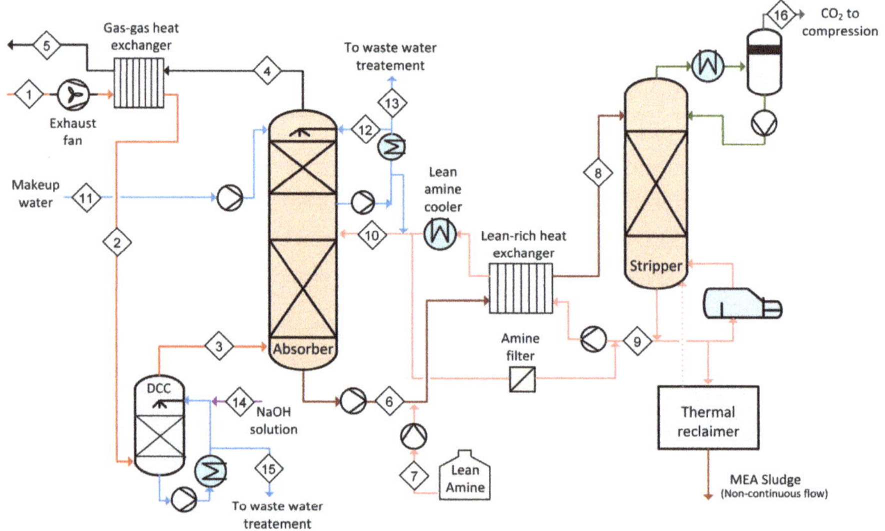

Figure 9. Process flowsheet of the MEA CO_2 capture process.

Table 7. Key process data for the MEA process.

	First $\frac{1}{2}$ Year	Second $\frac{1}{2}$ Year
Air leak	Low	Medium
CO_2 avoided from flue gas, AC_{fg} (%)	90	90
Captured and conditioned CO_2 mole fraction (wet basis)	0.999	0.998
Lean CO_2 loading (mol/mol MEA)	0.22	0.22
Rich CO_2 loading (mol/mol MEA)	0.5	0.5
Specific reboiler duty CO_2 desorber (MJ/kg CO_2)	3.76	3.80
Reboiler temperature CO_2 desorber (°C)	128.4	128.4
Steam generation from waste heat (%)	4%	4%
Steam generation in NG boiler (%)	96%	96%
Specific power consumption (MJ/kg CO_2)	0.521	0.549
Cooling demand (MJ/kg CO_2)	3.80	3.63
MEA make-up (kg/kg CO_2)	0.001	0.001
Additional ammonia solution for NO_x removal (kg/kg CO_2)	0.0002	0.0002
NaOH solution for SO_x removal (kg/kg CO_2)	0.001	0.001
Process water make-up (kg/kg CO_2)	0.473	0.528

The regenerator reboiler requires 96–97 MW steam, corresponding to 3.76–3.80 MJ/kg CO_2, at approximately 128 °C. Out of this, 3.7 MW can be covered by heat recovery at the cement plant (Figure 2), while the rest must be covered by a NG fired boiler.

The concentration of NO_x at the absorber inlet should be limited to 410 mg/Nm^3 for MEA-based CO_2 absorption [33]. An increase in NO_x reduction rate to reach the acceptable level is achieved by increasing the injection rate of ammonia solution in the SNCR system, with 1.5 times the stochiometric amount of the additional NO_x to be reduced. The concentration of SO_x at the absorber inlet should be limited to 10 ppm$_v$ prior to the MEA capture process. The SO_x level is reduced by injection of 50% NaOH solution in the DCC. It is assumed that SO_x is selectively removed by the stochiometric amount of NaOH.

Thermal reclaiming of MEA solvent for removal of amine degradation products is not included in the process model. In this sub-process a slipstream with amine solution is vaporized, the vapor is returned to the main process, while the less volatile degradation products are removed from the process. The amount of degradation products, the energy requirement of the reclaimer, and the amount of solvent lost in the reclaimer waste stream are estimated based on the studies by IEAGHG [34] and Knudsen et al. [35].

The captured CO_2 is assumed to be conditioned for pipeline transport by three stage compression up to 80 bar with triethylene glycol (TEG) dehydration and pumping to 110 bar.

4.3. Oxyfuel

The core oxyfuel cement process is simulated with VDZs in-house cement process model. A process flowsheet of the plant model is shown in Figure 10, and resulting key process data are presented in Table 8. The model is based on a model of the reference cement plant [29,30]. It has been adapted to the oxyfuel process as part of the work in the ECRA CCS project [36–39], and model parameters are tuned based on experimental work in CEMCAP [19,20].

The CPU is simulated using Aspen HYSYS V9, as a single stage flash self-refrigerated unit. The power consumption of the CPU is 0.4 MJ/kg CO_2. The heat required for dehydration is assumed to be 21.9 kJ/kg CO_2 (estimated based on [40]) and provided by electric heaters. The waste heat recovery system is designed and simulated using the Sequential Framework for heat exchanger network synthesis [41]. The heat required for ASU dehydration is assumed to be provided by electric heaters.

The power demand related to other utilities at the plant are slightly changed compared to the reference kiln (*cf.* Table 3), due to changed gas compositions and flow rates in some existing fans, and the added power demand of new fans. This gives a small increase in power consumption of cement plant utilities from 132 kWh/t_{clk} to 139 kWh/t_{clk} [42].

Details about the design and operating parameters of the waste heat recovery system and the CPU are described by Jamali et al. [20].

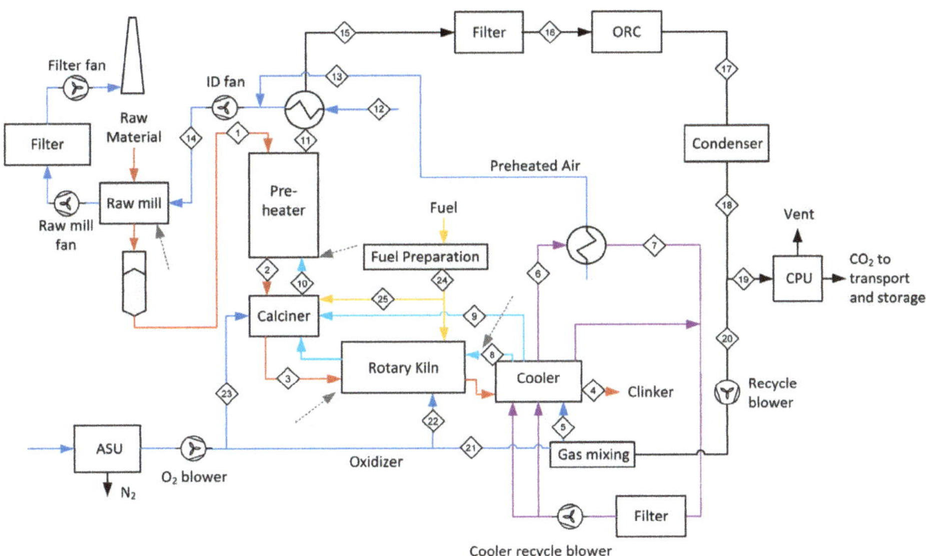

Figure 10. Process flowsheet of the oxyfuel plant.

Table 8. Key process data for the oxyfuel process.

Parameter	Value
False air at preheater outlet (vol%)	6.3
Clinker production (t/h)	125
CO_2 avoided from flue gas, AC_{fg} (%)	90
Captured and conditioned CO_2 mole fraction (wet basis)	0.973
Coal consumption (MW)	109
O_2 flow rate (t/h)	31.0
Generated CO_2 (t/h)	110
Captured CO_2 to storage (t/h)	99
Cooling (MW)	47.2
Net power consumption (MW)	35.1
ASU (MW)	*7.5*
CPU (MW)	*12.1*
Cooling system (MW)	*0.9*
Other utilities (MW)	*17.4*
Power generation in ORC (MW)	*−2.9*

4.4. Chilled Ammonia Process

The chilled ammonia process is simulated in Aspen Plus V8.6 using a rate-based model developed with experimental data from pilot tests [18]. These pilot plant tests, which were performed with synthetic cement plant flue gas streams, are additionally used to verify the final process simulations [34]. A process flowsheet of the system is shown in Figure 11 and resulting key process data are presented in Table 9.

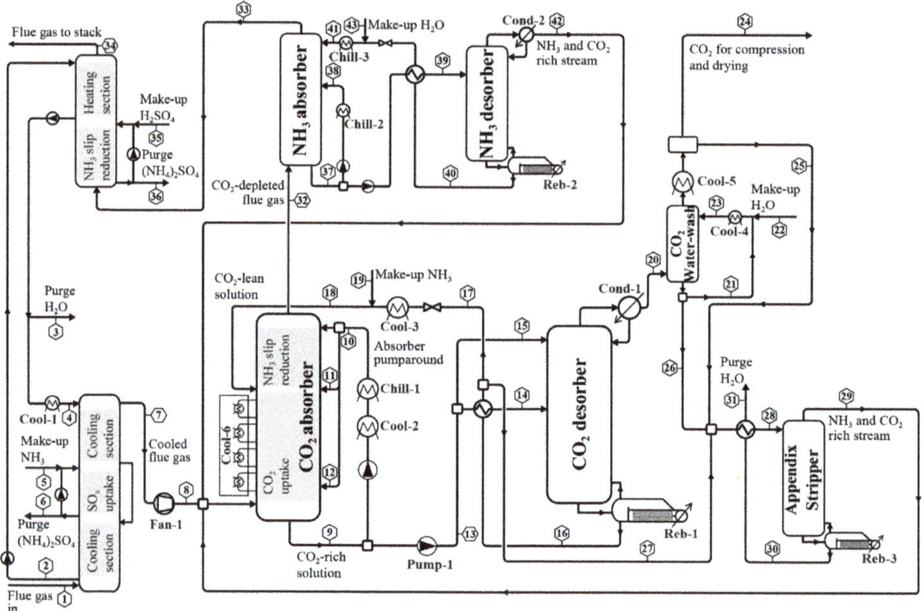

Figure 11. Process flowsheet of the chilled ammonia process.

Table 9. Key process data for the chilled ammonia process.

	First $\frac{1}{2}$ Year	Second $\frac{1}{2}$ Year
Air leak	Low	Medium
CO_2 avoided from flue gas, AC_{fg} (%)	90	90
Captured and conditioned CO_2 mole fraction (wet basis)	0.999	0.999
NH3 concentration CO_2-lean stream (mol/kg H_2O)	5	6
CO_2 loading CO_2-lean stream (mol/mol NH3)	0.415	0.387
CO_2-lean to inlet flue gas flowrate ratio (kg/kg)	9.0	6.5
Pumparound split fraction (%)	0.12	0.18
Pumparound temperature (°C)	12	9
CO_2 desorber pressure (bar)	25	25
Cold-rich split fraction (%)	0.025	0.035
Specific reboiler duty CO_2 desorber (MJ/kg CO_2)	2.12	2.20
Reboiler temperature CO_2 desorber (°C)	155.5	157.0
Specific reboiler duty NH3 desorber (MJ/kg CO_2)	0.13	0.18
Reboiler temperature NH3 desorber (°C)	108.8	108.8
Specific reboiler duty appendix stripper (MJ/kg CO_2)	0.05	0.07
Reboiler temperature appendix stripper (°C)	115.0	115.0
Steam generation from waste heat (%)	8	7
Steam generation in NG boiler (%)	92	93
Specific power consumption (MJ/kg CO_2)	0.315	0.338
Cooling water demand (MJ/kg CO_2)	3.23	3.40
Process water make-up (kg/kg CO_2)	0.009	0.014
Solvent makeup (kg/kg CO_2)	0.002	0.003
H_2SO_4 consumption (g/kg CO_2)	0.6	1.0

Waste heat from the cement plant (Figure 2) is assumed to cover all the heat required for the reboiler in the NH_3 desorber and the appendix stripper (steam at ~110 °C), and a small part of the remaining heat demand (steam at ~145 °C for the CO_2 desorber). Refrigeration of the CO_2 absorber pumparound and of the water stream entering the top of the NH_3 absorber to 12–13 °C and to 15 °C, respectively, is required. For estimation of power consumption and cooling demand of the refrigeration system, coefficients of performance of 7–8 are assumed [15]. Since the captured CO_2 leaves the CO_2 desorber at elevated pressure—i.e., 25 bar which is optimal for minimization of SPECCA—only one compression stage in addition to the final pumping is assumed in the compression process.

A detailed description of the simulation and optimisation procedure of the core process is described in the original work by Pérez-Calvo et al. [43].

4.5. Membrane-Assisted CO_2 Liquefaction

The membrane-assisted CO_2 liquefaction system is simulated with Aspen HYSYS V9, with the Peng–Robinson equation of state. A multicomponent membrane model that has been integrated into the HYSYS interface is used to simulate the membrane unit. A process flowsheet of the system is shown in Figure 12 and resulting key process data are presented in Table 10.

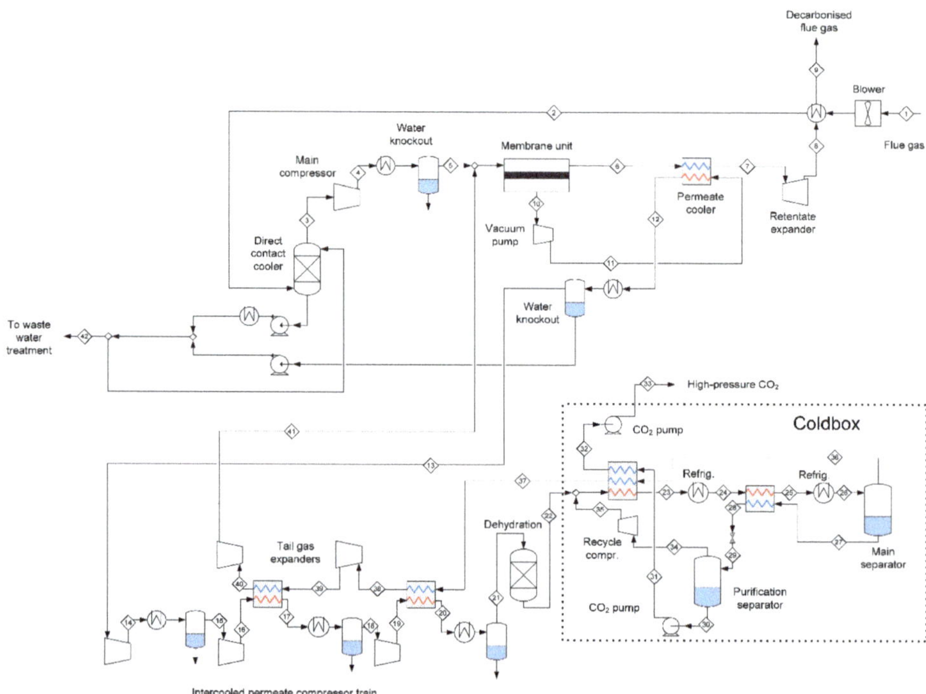

Figure 12. Process flowsheet of the membrane-assisted CO_2 liquefaction process.

Table 10. Key process data for membrane-assisted CO_2 liquefaction system.

	First $\frac{1}{2}$ Year	Second $\frac{1}{2}$ Year
Air leak	Low	Medium
Membrane area (m^2)	228,000	228,000
Membrane feed pressure (bar)	2.23	2.50
Main separator pressure (bar)	29	32
CO_2 avoided from flue gas, AC_{fg} (%)	90	90
Captured and conditioned CO_2 mole fraction (wet basis) (%)	99.40	99.27
Specific power consumption (MJ/kg CO_2)	1.23	1.44
Cooling water demand (MJ/kg CO_2)	1.54	1.51
Additional ammonia solution for NO_x removal (kg/kg CO_2)	0.0002	0.0002
NaOH solution for SO_x removal (kg/kg CO_2)	0.001	0.001

Membrane permeance data representative for the membrane that was tested within CEMCAP are used in the model (Table 11). It should be noted that the selection of membranes available for testing was limited.

Table 11. Membrane permeance and CO_2 selectivity for relevant components.

Component	Permeance (m3STP/m2bar-h)	CO_2 Selectivity over Component
CO_2	2.7	1
N_2	0.054	50
O_2	0.216	12.5
H_2O	54	0.05

The need for NO_x and SO_x removal from the flue gas stream beyond what is already required to not exceed the permitted limits for pollutant emissions depends on the tolerance of the membrane material. It is assumed that NO_x and SO_x removal is performed the same way and to the same extent as for the MEA absorption process.

A more detailed description of the process design and simulation is given by Berstad et al. [44].

4.6. Calcium Looping—Tail-End Configuration

The tail-end calcium looping process and its integration with the reference cement kiln are modeled with PoliMi's in-house process simulator GS [31]. A process flowsheet of the system is shown in Figure 13 and resulting key process data are presented in Table 12.

The circulating fluidized bed (CFB) reactor model presented by Romano [45] is used to simulate the carbonator, which includes the carbonation kinetic expression proposed by Grasa et al. [46]. The calciner is modeled assuming complete calcination of the sorbent. Based on the data presented by Martínez et al. [47], this assumption is justified considering the high residence time of the solids in a CFB calciner and the assumed calcination temperature of 920 °C. Further details about simulations of the CaL reactors are given by Spinelli et al. [48], and details on the methodology used for simulation of the overall process, with extensive sensitivity analysis on the main process parameters are given by De Lena et al. [49].

The heat recovery steam cycle parameters are set according to the thermal input available as defined in the CEMCAP framework [15]. The CPU is simulated using Aspen HYSYS V9, as a single stage flash self-refrigerated unit, slightly modified compared to the CPU of the oxyfuel process. The resulting power consumption of the CPU is 0.4 MJ/kg CO_2. The heat required for dehydration in the CPU is assumed to be 16.6 MJ/kg H_2O (estimated based on Kemper et al. [40]). The dehydration heat required in the CPU and the ASU are assumed to be provided by some of the steam generated in the process.

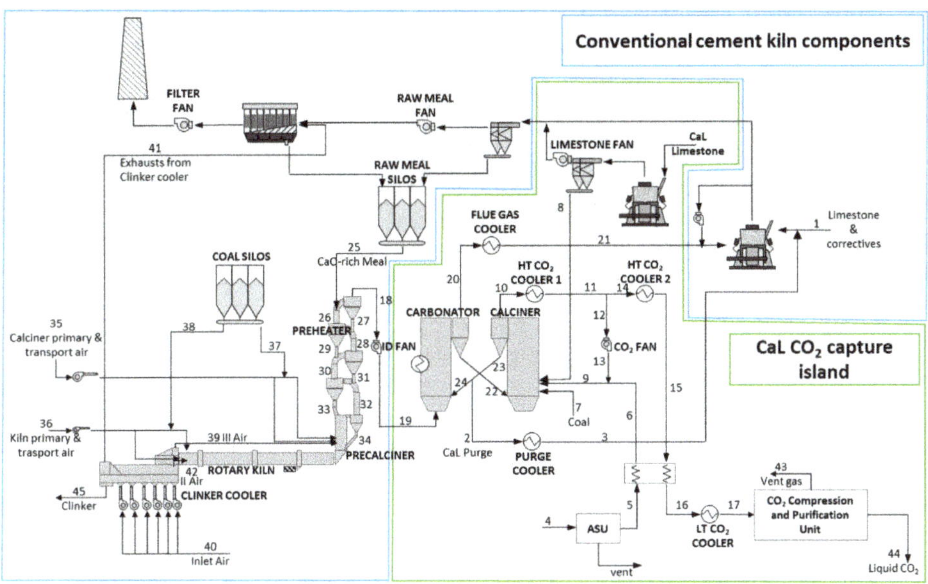

Figure 13. Process flowsheet of the tail-end calcium looping process integrated with the reference cement kiln.

Table 12. Key process data for the reference cement kiln with tail-end calcium looping CO_2 capture.

Parameter	Value
Integration level (%)	50
F_0/F_{CO2}	0.60
F_{Ca}/F_{CO2}	2.94
Clinker production (t/h)	117.7
CO_2 avoided from flue gas, AC_{fg} (%)	91
Coal consumption—rotary kiln (MW)	39.9
Coal consumption—calciner (MW)	27.6
Coal consumption—CaL calciner (MW)	164.6
O_2 flow rate (t/h)	51.5
CO_2 capture efficiency of carbonator (%)	90.0
CO_2 generated from fuel combustion (t/h)	77.8
CO_2 generated from raw meal calcination (t/h)	67.3
CO_2 emission from kiln flue gas (t/h)	5.8
CO_2 emission from the CPU vent gas (t/h)	3.6
Captured CO_2 to storage (t/h)	136.1
Captured and conditioned CO_2 mole fraction (wet basis) (%)	96
Cooling [1] (MW)	33.3
Net power consumption (MW)	6.79
ASU (MW)	*11.64*
Fans in CaL system (MW)	*2.70*
CPU (MW)	*16.11*
Cooling system (MW)	*0.67*
Auxiliaries for calciner fuel grinding (MW)	*0.18*
Cement plant auxiliaries (MW)	*15.49*
Power generation in steam cycle (MW)	*−40.00*

[1] Cooling of steam cycle not included.

4.7. Calcium Looping—Integrated Configuration

As for the tail-end CaL technology, the process evaluation is based on simulations performed with Polimi's in-house process simulation code GS [31]. A process flowsheet of the system and its integration with the reference plant is shown in Figure 14 and resulting key process data are summarized in Table 13.

A one-dimensional, steady-state model, which is described by Spinelli et al. [48,50], has been used for the calculation of the entrained flow carbonator. Sorbent conversion kinetics are described by the random pore model proposed by Grasa et al. [51]. For the calciner, an outlet temperature of 920 °C has been assumed to calculate the heat input needed in that reactor to heat up and calcine the recarbonated raw meal from the carbonator. Further details about simulations of the cement kiln with the integrated CaL configuration are given by De Lena et al. [52].

For the heat recovery steam cycle, the CPU and the ASU the same simulation approaches and assumptions as for the CaL tail-end process are used.

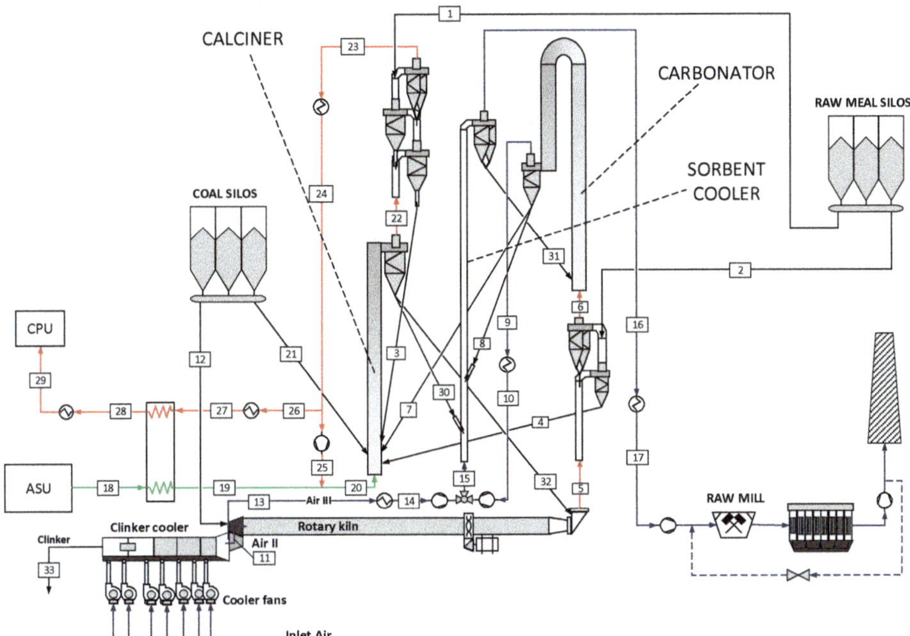

Figure 14. Process flowsheet of the reference cement kiln with integrated calcium looping CO_2 capture.

Table 13. Key process data for the reference cement kiln with integrated calcium looping CO_2 capture.

Parameter	Value
F_0 / F_{CO2}	3.96
F_{Ca} / F_{CO2}	3.90
Clinker production (t/h)	117.4
CO_2 avoided from flue gas, AC_{fg} (%)	93
Coal consumption—rotary kiln (MW)	37.5
Coal consumption—CaL calciner (MW)	139.8
O_2 flow rate (t/h)	44.1
CO_2 capture efficiency of carbonator (%)	82.0
CO_2 generated from fuel combustion (t/h)	59.4
CO_2 generated from raw meal calcination (t/h)	64.4
CO_2 emissions at stack from kiln flue gas (t/h)	2.9
CO_2 emissions at stack from the CPU vent gas (t/h)	3.6
Captured CO_2 to storage (t/h)	117.0
Captured and conditioned CO_2 mole fraction (wet basis) (%)	96
Cooling [1] (MW)	29.1
Net power consumption (MW)	20.41
ASU (MW)	*9.96*
Fans in CaL system (MW)	*1.21*
CPU (MW)	*14.11*
Cooling system (MW)	*0.58*
Auxiliaries for calciner fuel grinding (MW)	*0.26*
Cement plant auxiliaries (MW)	*15.36*
Power generation in steam cycle (MW)	*−21.06*

[1] Cooling of steam cycle not included.

5. Comparative Technical Evaluation of the CO_2 Capture Technologies

5.1. Emission Abatement

The CO_2 capture ratio, the CO_2 avoided from the flue gas, and the equivalent CO_2 avoided evaluate the effect that the implementation of a capture technology has on the CO_2 emissions of a plant. These KPIs calculated for the investigated technologies and the reference technology are given in Table 14.

Table 14. Direct and indirect emissions, CO_2 capture ratio, and CO_2 avoided.

	Ref. Plant VDZ	Ref. Plant PoliMi	MEA	Oxy-Fuel	CAP	MAL	CaL Tail-end	CaL Int.
Direct CO_2 emissions at stack (kg/t_{clk})	846	865	84	88	83	84	78	57
Direct CO_2 emissions due to steam generation (kg/t_{clk})	0	0	172	0	104	0	0	0
Indirect CO_2 emissions (kg/t_{clk})	34	34	64	74	53	109	15	46
CO_2 capture ratio (CCR) (%)	-	-	90	90	90	90	94	95
CO_2 avoided from flue gas (AC_{fg}) (%)	-	-	90	90	90	90	91	93
Equivalent CO_2 avoided (AC_{eq}) (%)	-	-	64	82	73	78	90	89

The CCR of all technologies range between 90% and 95%, and the CO_2 avoided from the flue gas range between 90 and 93%. For the MEA, CAP, and MAL technologies, the CO_2 avoided from flue gas is by definition equal to the CCR, since there is no change in internal fuel combustion generating CO_2 within the kiln or the capture process when these technologies are installed. For the oxyfuel technology the value is approximately the same as the CCR—the specific fuel consumption changes in the oxyfuel technology compared to the standard reference kiln, but the change is very small. For the calcium looping technology, the CO_2 avoided from the flue gas is lower than the CCR, because the capture of CO_2 generated by fuel combustion within the calciner is not counted as CO_2 avoided.

The equivalent CO_2 avoided of all technologies range between 64% and 90%. With this KPI all direct emissions at the plant, as well as indirect emissions related to power consumption or generation are taken into account. This value is lower than the CO_2 avoided for all technologies due to the contribution of direct emissions from steam generation and indirect emissions associated to power import.

All the investigated technologies have equivalent CO_2 avoided in the range 73–90%, which is higher than the reference technology MEA with 64%. The CaL technologies end up with the highest equivalent CO_2 avoided. These technologies have no additional direct emissions as most of the CO_2 from the additional coal combustion is captured, and negative or low added indirect CO_2 emissions thanks to the internal power production.

5.2. Energy Performance

The energy inputs of the reference cement plant without and with CO_2 capture are in the form of coal, natural gas, and power. For some CO_2 capture technologies power is also generated on-site. A summary of the energy inputs is given in Table 15.

Table 15. Reference plant energy input without and with CO_2 capture technologies.

	Ref. Plant (VDZ)	Ref. Plant (PoliMi)	MEA	Oxy-Fuel	CAP	MAL	CaL Tail-End	CaL Int.
Coal consumption (kJ/kg_{clk})	3135	3241	3135	3139	3135	3135	7100	5436
NG consumption (kJ/kg_{clk})	0	0	3073	0	1859	0	0	0
Power consumption (kJ/kg_{clk})	474	474	881	1094	723	1491	1431	1272
Power generation (kJ/kg_{clk})	0	0	0	−83	0	0	−1,223	−646

The specific primary energy consumption for CO_2 avoided of the technologies are presented in Figure 15 and Table 16. All the investigated technologies have clearly lower SPECCA values than the reference technology. The oxyfuel technology has a SPECCA of 1.63 MJ/kg CO_2, which is the lowest value among the SPECCA values of the investigated technologies. The chilled ammonia and membrane-assisted liquefaction technologies have SPECCA values of 3.75 and 3.22 MJ/kg CO_2 respectively, while the calcium looping tail-end and integrated technologies have SPECCA values of 4.07 and 3.17 MJ/kg CO_2.

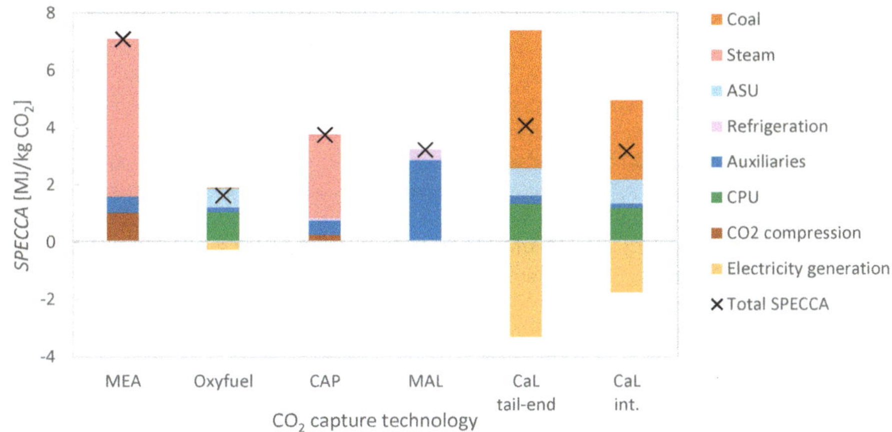

Figure 15. Specific primary energy consumption for CO_2 avoided (SPECCA).

Table 16. Break-down of specific primary energy consumption for CO_2 avoided (SPECCA).

	MEA	Oxy-fuel	CAP	MAL	CaL Tail-End	CaL Int.
Added equivalent specific primary energy consumption (MJ/t$_{clk}$)	**3959**	**1173**	**2401**	**2216**	**3280**	**2528**
Coal consumption	0	4	0	0	3859	2195
NG consumption for steam generation	3073	0	1859	0	0	0
Electric power consumption	887	1351	542	2216	2086	1740
ASU	*0*	*473*	*0*	*0*	*776*	*666*
Refrigeration	*0*	*0*	*59*	*260*	*0*	*0*
Auxiliaries [1]	*312*	*119*	*316*	*1955*	*236*	*132*
CPU	*0*	*759*	*0*	*0*	*1074*	*943*
CO$_2$ compression	*575*	*0*	*167*	*0*	*0*	*0*
Electric power generation	0	−182	0	0	−2665	−1408
Equivalent specific CO$_2$ avoided (kg/t$_{clk}$)	**559**	**719**	**640**	**687**	**806**	**797**
At cement kiln stack	761	758	762	761	787	808
Steam consumption (NG fired boiler)	−172	0	−104	0	0	0
Electric power consumption	−30	−45	−18	−74	−70	−58
ASU	*0*	*−16*	*0*	*0*	*−26*	*−22*
Refrigeration	*0*	*0*	*−2*	*−9*	*0*	*0*
Auxiliaries [1]	*−10*	*−4*	*−11*	*−65*	*−8*	*−4*
CPU	*0*	*−25*	*0*	*0*	*−36*	*−31*
CO$_2$ compression	*−19*	*0*	*−6*	*0*	*0*	*0*
Electric power generation	0	6	0	0	89	47
SPECCA (MJ/kg CO$_2$)	**7.08**	**1.63**	**3.75**	**3.22**	**4.07**	**3.17**

[1] Pumps, fans, fuel grinding (CaL), compressors, thermal reclaimer (MEA), dehydration (MAL) and cooling water system (cooling tower and pumps).

The most important contributions to the SPECCA differ among the technologies. For the MEA technology the primary energy consumption related to the steam required in the process is responsible for the largest part of the added equivalent primary energy consumption and reduction in equivalent CO_2 avoided. For the oxyfuel technology, the added equivalent primary energy consumption and reduction in equivalent CO_2 avoided are almost entirely due to the increased electric power consumption. The CPU is the largest power consumer, followed by the ASU and the fans. Electric power generation from waste heat reduces the net power consumption by almost one fifth. For the chilled ammonia process, the steam consumption makes up the largest part of the primary energy consumption and reduction in equivalent CO_2 avoided. The steam consumption makes up around three-quarters of these values, while the electric power consumption is responsible for the rest. For the membrane-assisted CO_2 liquefaction process, electric power consumption is responsible for all added equivalent primary energy consumption and reduction in equivalent CO_2 avoided, where around four-fifths are due to fan, pump and compressor work in the process, and the rest is mainly due to the refrigeration system. For both calcium looping processes, coal consumption, electric power consumption and electric power generation are important for the final SPECCA value. The considerable electric power generation is especially important for the tail-end technology as it contributes to a reduction in both added equivalent specific primary energy consumption and equivalent specific CO_2 avoided. This essentially means that the electricity generated covers a part of the cement plant's demand as well as the demand of the CO_2 capture process, resulting in lower electric power consumption per unit of clinker produced.

The characteristics of the power generation system in terms of electricity generation efficiency, η_{el}, and the specific CO_2 emissions of the electricity generation, e_{el}, have an impact on the SPECCA. The generation efficiency and specific CO_2 emissions are directly linked to the power generation technology that is assumed to provide the electricity required by the processes.

To investigate the impact of the values of η_{el} and e_{el}, SPECCA values are calculated with several different options for power generation. The average electricity mix in EU-28 in 2014 is used as basis in the calculations. The alternative cases are summarized in Table 4, and the results are shown in Figure 16. For the calcium looping tail-end technology, the SPECCA increases with increasing electricity generation efficiency and decreasing specific CO_2 emissions, while the opposite is observed for all the other CO_2 capture technologies. This is because the CaL tail-end technology generates enough electricity to cover both its own demand and a part of the electricity demand of the cement plant, effectively substituting some of the electricity that was bought from the grid in the reference cement plant. With increasing generation efficiency of the power system and a decrease in the associated specific CO_2 emissions, the reduction in indirect added equivalent specific primary energy consumption becomes smaller, as well as the indirect avoided equivalent specific CO_2 emissions.

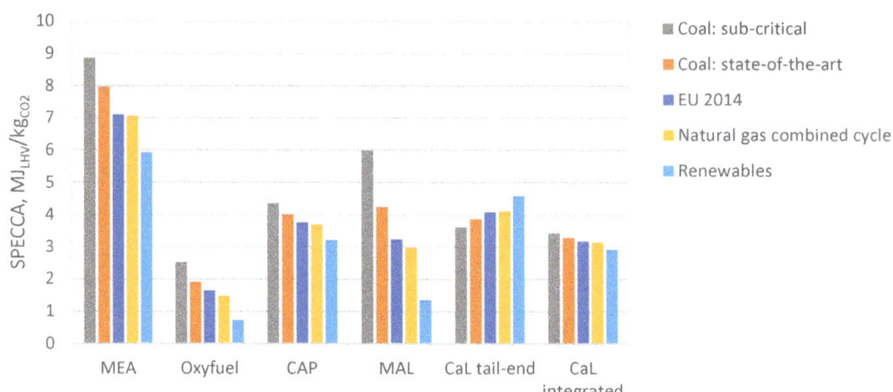

Figure 16. Specific primary energy consumption for CO_2 avoided (SPECCA) of the technologies with different power generation scenarios.

For oxyfuel and MAL technologies, where the main added energy input is in the form of electric power, the SPECCA value is highly dependent on the characteristics of the power generation system. In the case of electricity being solely generated from renewables, the SPECCA is reduced by more than half compared to the EU 2014. On the other hand, the SPECCA of the MAL technology is almost doubled in the worst case of electricity generation from sub-critical pulverized coal plants. The SPECCA values of the MEA and CAP technologies are also significantly affected by the different power generation cases.

As mentioned, the generation of steam is responsible for the largest part of the added primary energy consumption and equivalent CO_2 emissions for the absorption-based technologies, MEA and CAP. The SPECCA is therefore largely dependent on amount of waste heat available at the plant, and the selected strategy for steam supply. One alternative to steam generation in NG fired boilers is steam import from a coal fired combined heat and power (CHP) plant. The associated primary energy consumption can be assumed to be 0.34-0.68 MW_{th}/MW_{th} and the associated CO_2 emissions 116-231 kg/MWh_{th} for steam temperatures in the range 100-160 °C, considering a conversion efficiency for coal to power without steam extraction of 39.1% and a conversion efficiency for steam to power of 13.3-26.6% for steam in this temperature range [24]. The energy consumption and CO_2 emissions of the CHP plant are partly assigned to the generated steam and partly to the generated electric power. The SPECCA is reduced with 47% for MEA and 35% for CAP in this case, but it should be mentioned that few cement plants are located close to a power plant.

It is assumed in this study that cooling water is supplied by a cooling tower, that requires power for fans and water pumping, and delivers cooling water at 18 °C. If the cooling could be performed with water from the sea or a river, less power consumption would be required by the cooling system itself, and the cooling water could in many cases be delivered at lower temperature levels. This would benefit all the technologies slightly.

5.3. Retrofitability

The overall retrofitability assessment of the technologies is shown in Table 17, and more detailed reasoning is given in the following subsections. In general, it can be noted that the post-combustion technologies are easier to retrofit, while the integrated technologies are more challenging.

Table 17. Assessment of retrofitability.

Indicator	MEA	Oxyfuel	CAP	MAL	CaL Tail-End	CaL Int.
Impact on the cement production process	[green]	[orange]	[green]	[green]	[green]	[yellow]
Equipment and footprint	[yellow]	[orange]	[yellow]	[yellow]	[yellow]	[orange]
Utilities and services	[yellow]	[yellow]	[yellow]	[yellow]	[yellow]	[yellow]
Introduction of new chemicals/subsystems	[yellow]	[yellow]	[yellow]	[green]	[yellow]	[yellow]
Available operational experiences	[green]	[orange]	[yellow]	[orange]	[yellow]	[orange]

5.3.1. Impact on the Cement Production Process

The application of post-combustion technologies does not affect the actual clinker burning process or the clinker quality, as they can be installed as independent units that flue gas is sent to before the stack. During the construction phase, only a short stop of the clinker production would be required for the rerouting of the flue gas. This rerouting can be performed during the yearly maintenance period when the plant is shut down and does therefore not need to affect the operability of the plant. The tail-end CaL process is slightly integrated with the cement kiln since sorbent purge is ground and used as raw meal in the kiln. However, this does not include any risk for the plant operability or the clinker quality. Therefore, all post-combustion technologies are marked green.

The oxyfuel and the integrated CaL processes require significant modification of the production process. As a result, potential operational problems due to the capture technology can directly affect the operability of the plant. There is also an increased risk related to the quality of the produced clinker [53]. In theory changes of the gas atmosphere, of gas temperatures, and other process conditions can be managed so that optimum clinker production still can be achieved, but this remains to be proven. A long production stop is required during the construction phase for both these technologies.

In the oxyfuel process, the clinker cooler, rotary kiln, calciner, and preheater are modified, whereas for the integrated CaL technology only the calciner and preheater are changed, while the cooler and the kiln are unchanged. The oxyfuel process is therefore marked orange and the integrated CaL process is marked yellow.

5.3.2. Equipment and Footprint

Every capture process will need installation of some additional equipment which requires available space, and therefore at least some attention is required regarding this aspect for all technologies. The most important difference between the technologies is whether the equipment must be installed close to the kiln line or if it can be installed further away.

For all the post-combustion technologies the required equipment can be installed anywhere at the plant, and there is also some flexibility for splitting up the systems and installing different units at different locations at the plant, so these technologies are marked yellow. The oxyfuel and the integrated CaL processes are integrated with the kiln system itself, and these technologies require space close to the kiln line. These technologies are therefore marked as orange.

5.3.3. Utilities and Services

The need for utilities and services of the technologies is summarized in Table 18. The two solvent based processes MEA and CAP require considerable amounts of steam and also some power, in addition to the solvents MEA and ammonia. Each of these points should be possible to handle in most cases, but still require some attention. These technologies are therefore marked yellow.

Table 18. Additional utilities and services required for the reference cement plant.

	MEA	Oxyfuel	CAP	MAL	CaL Tail-End	CaL Int.
Electric power	14 MW	19 MW	8 MW	34 MW	−9 MW	5 MW
Steam	96 MW	-	61 MW	-	-	
Coal	-	-	-	-	126 MW	71 MW
Solvent	MEA	-	ammonia	-	-	-
Raw meal quality	-		-	-	15–20% Ca as limestone	-

For the oxyfuel and MAL processes, only additional electric power is required, but in both cases the magnitude of the power demand is considerable. These technologies are marked yellow, because some attention is needed on the local power grid capacity.

For the CaL processes, additional coal is required. Additional electric power is also needed, but on the other hand power is also generated from the waste heat. Depending on the integration level, the net power consumption at the plant can be positive or negative. If there is power export from the plant, infrastructure would be required for this. If the power generation is balanced with the power consumption, there would not be any need for import of power, which is an advantage. For the tail-end configuration 15–20% of the calcium must be fed to the plant as limestone. Some attention is needed for the plant retrofit, with respect to coal import, import/export of power, and raw meal quality. The two CaL processes are therefore marked yellow.

5.3.4. Introduction of New Chemicals/Subsystems

The operation of the MAL requires the installation of a refrigeration system, which implies that refrigerants will be present at the plant, and small amounts of NaOH for SO_x removal. This should be relatively easy to handle, and therefore this technology is marked in green.

All capture technologies which require the production and use of oxygen, a chemical that can increase the risk for fires and explosions at the plant, could require a more complex permitting process and would require that new procedures and routines are established. However, the use of oxygen is normal in many industries, so although attention is needed it can still be handled. Therefore, the oxyfuel and the CaL process are marked yellow.

The MEA process and the CAP require aqueous solutions of amines or ammonia as solvent. Amines and their degradation products, as well as ammonia, are poisonous and dangerous for the environment. The use of these chemicals requires a permitting process, whose complexity depends on national/local regulations. Furthermore, new procedures and routines must be established to ensure safe operation at the plant. Ammonia is already commonly used at cement plants for NO_x removal systems, but in very low quantities compared to what is required for the CAP. As for oxygen, significant experience is available from other industries, so MEA and CAP are marked yellow.

5.3.5. Available Operational Experiences

The MEA process is the most mature capture technology and a lot of information is already available. Other types of amines have been tested for flue gas treatment from an operational plant, and it can be expected that a retrofit should be possible without major problems. The most advanced testing done so far with amines on cement flue gas so far are the pilot trials with Aker Solutions mobile test unit, which is a fully integrated prototype of the system, for six months in an operational kiln at Norcem in Brevik, Norway. The testing at Norcem showed good stability of the solvent towards cement flue gas, and no technical show-stopper was identified. A capture ratio of 90% was obtained, and 370 tonnes of CO_2 were captured over 2700 h [54]. Furthermore, there is a lot of operational experience with amine absorption from the demonstration for coal power plants in commercial scale at Boundary Dam. Consequently, this technology is marked green.

For the oxyfuel process, burner, calciner, and clinker cooler pilot trials have been conducted in industrial relevant environment as a part of the CEMCAP project. A 500 kW oxyfuel cement kiln burner prototype has been successfully demonstrated at the University of Stuttgart [19,55]. Oxyfuel calcination has been demonstrated in a 50 kW reactor, also at the University of Stuttgart [56]. A clinker cooler pilot has been successfully operated with clinker directly from a real industrial kiln line at HeidelbergCement's plant site in Hannover [57]. However, the full system has not been operated as a whole yet, and experience from the power sector is not directly transferrable to the cement sector, so the technology is marked orange.

The chilled ammonia process has been demonstrated for flue gas concentrations ranging from typical natural gas-fired power plants to typical coal-fired power plants in several pilots including the 58 MW$_{th}$ AEP Mountaineer pilot [58] and the 40 MW$_{th}$ TCM pilot [59]. In CEMCAP, the chilled ammonia absorber, direct contact cooler (DCC), and water wash units were tested for the cement application, since these are the units that are affected by a change in the flue gas conditions. All three units were tested in GE's 1 tonne CO_2 per day pilot facility in Vaxjö, verifying that the process can be applied in the cement industry [44]. Due to the experience from the pilot plants in the power sector, no major risks are foreseen for the operation of all units together or for the scale-up of the system. Based on this, the technology is marked yellow.

For the membrane-assisted CO_2 liquefaction process, the liquefaction system and membranes have been tested separately. The liquefaction part of the process has been tested as a part of CEMCAP in a pilot facility at SINTEF Energy Research with liquefaction capacity of 10 tonne CO_2 per day, and a separation performance consistent with expectations based on vapour-liquid equilibrium data was demonstrated [21]. One type of membranes as tested in the lab for cement specific flue gas in the

CEMCAP project, but more advanced testing has been carried out for fixed-site-carrier membranes at the Norcem plant in Brevik with flue gas from an operational kiln [60]. The full process with membranes and the liquefaction system integrated has not been operated, so therefore the technology is marked orange.

The tail-end calcium looping technology has been demonstrated for coal at the "la Pereda" power plant in a 1.7 MW_{th} pilot [61]. Within CEMCAP the technology has been tested for cement flue gas in a 30 kW pilot at CSIC, and a 200 kW pilot at the University of Stuttgart [26]. The testing has shown that fundamental parameters, such as the carbonation rate constant, are consistent with those in systems already tested for coal power plants at a large scale. This technology is therefore marked yellow.

For the calcium looping integrated technology, calcium looping experiments with entrained flow reactors have been carried out in the 30 kW pilot at CSIC. It has been shown on a laboratory scale that calcined materials with free CaO are able to adsorb CO_2 in an entrained flow environment. There is not yet any operational experience with the full system, so this technology is marked orange.

6. Conclusions

A consistent technical assessment of the oxyfuel process, the chilled ammonia process, membrane-assisted liquefaction, and the calcium looping technology with tail-end and integrated configurations retrofitted to a reference cement plant is performed. The technologies are benchmarked against MEA absorption which is used as a reference technology.

All the investigated technologies perform better than the reference technology both in terms of emission abatement and primary energy consumption. The equivalent CO_2 avoided under the defined conditions are in the range 73–90% for the investigated technologies, while it is 64% for the reference technology. The calcium looping technologies have the highest emission abatement performance with 89–90% equivalent CO_2 avoided. For this technology, most of the CO_2 generated by the additional primary energy consumption are also captured. The other technologies entail additional emissions due to imported electricity or steam generation, which reduce the equivalent CO_2 avoided despite the fact that all technologies capture 90% of the cement plant's emissions. The SPECCA values of the investigated technologies are in the range 1.63–4.07 MJ/kg CO_2, compared to 7.08 MJ/kg CO_2 for the reference technology. The oxyfuel process has the lowest SPECCA, with 1.63 MJ/kg CO_2. This is explained by a significantly lower primary energy demand than the other technologies and a medium value for equivalent CO_2 avoided.

The post-combustion technologies, particularly the reference technology MEA, are assessed as easier to retrofit than the integrated technologies. Clear advantages of the post combustion technologies are the low impact on the cement production process and the flexibility in the placing of new equipment at the cement plant. The oxyfuel and integrated CaL technologies, which are more integrated with the cement plant, are assessed as more challenging, although no 'showstoppers' were identified for their installation in existing plants.

Due to the different performance of the technologies with respect to the different KPIs, it is not possible to identify one overall winner among them. A high equivalent CO_2 avoided and low SPECCA will always be desired but depending on the specific cement plant and the local conditions, other aspects may become dominant, such as space requirements, utility requirements in the light of available infrastructure, or available operational experience.

Furthermore, it is shown that for the technologies that require a considerable amount of electric power, the electricity mix has a large effect on the SPECCA. This is particularly important for the membrane-assisted liquefaction process which has a high electricity demand, but no other primary energy consumption. In a scenario with a large share of renewable electric energy—e.g., the Norwegian mix—this technology has the second lowest SPECCA among the technologies. On the contrary, in a scenario based on coal power, which is relevant for some Eastern Europe countries, this technology has the highest SPECCA after the reference technology. Similarly, it is highlighted that different assumptions on the availability of waste heat, co-generated steam, cooling water, etc. affect the performance of the different capture technologies to a varying extent.

Based on these findings, it can be concluded that CCS in the cement industry should be performed with a portfolio of capture technologies. For identification of the optimal capture technology for a specific plant, a case specific evaluation should be performed considering the local conditions and constraints. The results and the discussion of their sensitivity presented above provide robust indications on which technologies may be favored under certain conditions.

An economic evaluation of the technologies is presented in Part 2 of this paper series.

Supplementary Materials: The following are available online at http://www.mdpi.com/1996-1073/12/3/559/s1, Process flowsheets: Figures 1.1–7.1; Stream data: Tables 1.1–7.2.

Author Contributions: M.V. and S.O.G. assembled the data and calculated quantitative KPIs. E.D.L., J.-F.P.-C., A.J., D.B., C.F., and R.A. performed process simulations. H.H., M.V., D.S., M.M., and M.R. performed retrofitability analysis. R.A., S.R., M.R., M.V., and S.O.G. defined the methodology. G.C. validated the cement-specific technical base of the study. M.R., M.M., M.G., R.A., and K.J. supervised the work. All authors contributed to the overall analysis and this was coordinated by M.V. and S.O.G. M.V. wrote the paper; all authors contributed to the text.

Funding: This project has received funding from the European Union's Horizon 2020 Research and Innovation Programme under grant agreement no. 641185, and the Swiss State Secretariat for Education, Research and Innovation (SERI) under contract number 15.0160.

Conflicts of Interest: The authors declare no conflict of interest.

Abbreviations

ASU	air separation unit
BAT	best-available technologies
CaL	calcium looping
CAP	chilled ammonia process
CAPEX	capital costs
CCS	carbon capture and storage
CFB	circulating fluidized bed
CPU	CO_2 purification unit
DCC	direct contact cooler
ECRA	European Cement Research Academy
HSS	heat stable salts
IL	integration level
KPI	key performance indicator
MAL	membrane-assisted CO_2 liquefaction
MEA	monoethanolamine
NG	natural gas
OPEX	operating costs
ORC	organic Rankine cycle
PC	pulverized coal
SNCR	selective non-catalytic reduction
SRK	Soave–Redlich–Kwong
TEG	triethylene glycol

Nomenclature

η_{el}	electricity generation efficiency
AC_{eq}	equivalent CO_2 avoided
AC_{fg}	CO_2 avoided from flue gas
CCR	carbon capture ratio
e_{clk}	specific direct CO_2 emissions
$e_{clk,eq}$	specific equivalent CO_2 emissions
$e_{clk,eq,ref}$	specific equivalent CO_2 emissions in reference plant

$e_{clk,fg}$	specific CO_2 emissions with kiln flue gas
$e_{clk,fg,ref}$	specific CO_2 emissions with kiln flue gas in reference plant
e_{el}	CO_2 emissions associated with electric power
$e_{el,clk}$	specific indirect CO_2 emissions
F_0	molar flowrate of fresh sorbent
F_{Ca}	molar flowrate of CaO entering carbonator
F_{CO2}	molar flowrate of CO_2
$\dot{m}_{clk}$	mass flowrate of clinker
$\dot{m}_{CO2,capt}$	mass flowrate of CO_2 captured
$\dot{m}_{CO2,gen}$	mass flowrate of CO_2 generated
$\dot{m}_{fuel}$	mass flowrate of fuel
$P_{el,clk}$	specific power consumption
q_{clk}	direct specific primary energy consumption
$q_{clk,\,eq}$	equivalent specific primary energy consumption
$q_{clk,eq,ref}$	equivalent specific primary energy consumption in reference plant
$q_{el,clk}$	indirect specific primary energy consumption
SPECCA	specific primary energy consumption for CO_2 avoided

References

1. International Energy Agency (IEA). *Technology Roadmap: Low-Carbon Transition in the Cement Industry*; IEA: Paris, France, 2018; Available online: https://webstore.iea.org/technology-roadmap-low-carbon-transition-in-the-cement-industry (accessed on 6 January 2019).
2. IEAGHG. *Deployment of CCS in the Cement Industry*; Report Number 2013/19; IEAGHG: Cheltenham, UK, 2013; Available online: https://ieaghg.org/docs/General_Docs/Reports/2013-19.pdf (accessed on 8 January 2019).
3. CSI (Cement Sustainability Initiative). In *CSI/ECRA-Technology Papers 2017: Development of State of the Art Technologies in Cement Manufacturing: Trying to Look Ahead*; World Business Council for Sustainable Development: Geneva, Switzerland, 2017; Available online: https://docs.wbcsd.org/2017/06/CSI_ECRA_Technology_Papers_2017.pdf (accessed on 6 January 2019).
4. Potocnik, J. 2013/163/EU: Commission Implementing Decision of 26 March 2013 establishing the best available techniques (BAT) conclusions under Directive 2010/75/EU of the European Parliament and of the Council on industrial emissions for the production of cement, lime and magnesium oxide. *Off. J. Eur. Union* **2013**, 1–45. Available online: http://data.europa.eu/eli/dec_impl/2013/163/oj (accessed on 6 January 2019).
5. Liang, X.; Li, J. Assessing the value of retrofitting cement plants for carbon capture: A case study of a cement plant in Guangdong, China. *Energy Convers. Manag.* **2012**, *64*, 454–465. [CrossRef]
6. Ozcan, D.C. Techno-Economic Study for the Calcium Looping Process for CO_2 Capture from Cement and Biomass Power Plants. Ph.D. Thesis, University of Edinburgh, Edinburgh, UK, 2014.
7. Jakobsen, J.; Roussanaly, S.; Anantharaman, R. A techno-economic case study of CO_2 capture, transport and storage chain from a cement plant in Norway. *J Clean Prod.* **2017**, *144*, 523–539. [CrossRef]
8. Gerbelová, H.; van der Spek, M.; Schakel, W. Feasibility Assessment of CO_2 Capture Retrofitted to an Existing Cement Plant: Post-combustion vs. Oxy-fuel Combustion Technology. *Energy Procedia* **2017**, *114*, 6141–6149. [CrossRef]
9. Roussanaly, S.; Fu, C.; Voldsund, M.; Anantharaman, R.; Spinelli, M.; Romano, M. Techno-economic Analysis of MEA CO_2 Capture from a Cement Kiln—Impact of Steam Supply Scenario. *Energy Procedia* **2017**, *114*, 6229–6239. [CrossRef]
10. Rodríguez, N.; Murillo, R.; Abanades, J.C. CO_2 Capture from Cement Plants Using Oxyfired Precalcination and/or Calcium Looping. *Environ. Sci. Technol.* **2012**, *46*, 2460–2466. [CrossRef] [PubMed]
11. Diego, M.E.; Arias, B.; Abanades, J.C. Analysis of a double calcium loop process configuration for CO_2 capture in cement plants. *J. Clean Prod.* **2016**, *117*, 110–121. [CrossRef]
12. Lindqvist, K.; Roussanaly, S.; Anantharaman, R. Multi-stage Membrane Processes for CO_2 Capture from Cement Industry. *Energy Procedia* **2014**, *63*, 6476–6483. [CrossRef]

13. Jordal, K.; Voldsund, M.; Størset, S.; Fleiger, K.; Ruppert, J.; Spörl, R.; Hornberger, M.; Cinti, G. CEMCAP—Making CO_2 Capture Retrofittable to Cement Plants. *Energy Procedia* **2017**, *114*, 6175–6180. [CrossRef]

14. Schorcht, F.; Kourti, I.; Scalet, B.M.; Roudier, S.; Sancho, L.D. Best Available Techniques (BAT) Reference Document for the Production of Cement, Lime and Magnesium Oxide. Industrial Emissions Directive 2010/75/EU. 2013. Available online: http://eippcb.jrc.ec.europa.eu/reference/BREF/CLM_Published_def.pdf (accessed on 8 January 2019).

15. Voldsund, M.; Anantharaman, R.; Berstad, D.; Cinti, G.; De Lena, E.; Gatti, M.; Gazzani, M.; Hoppe, H.; Martínez, I.; Monteiro, J.G.M.-S.; et al. *CEMCAP Framework for Comparative Techno-Economic Analysis of CO_2 Capture from Cement Plants (D3.2)*; 2018. [CrossRef]

16. Directive 2010/75/EU of the European Parliament and of the Council of 24 November 2010 on Industrial Emissions (Integrated Pollution Prevention and Control). 2010. Available online: https://eur-lex.europa.eu/legal-content/EN/TXT/?uri=celex%3A32010L0075 (accessed on 8 January 2019).

17. Norahim, N.; Yaisanga, P.; Faungnawakij, K.; Charinpanitkul, T.; Klaysom, C. Recent Membrane Developments for CO_2 Separation and Capture. *Chem. Eng. Technol.* **2018**, *41*, 211–223. [CrossRef]

18. Casillas, C.; Chan, K.; Fulton, D.; Kaschemekat, J.; Kniep, J.; Ly, J.; Merkel, T.; Nguyen, V.; Sun, Z.; Wang, X.; et al. Pilot Test Results from a Polaris^{TM} Membrane 1 MW_e CO_2 Capture System. In Proceedings of the Carbon Management Technology Conference, Houston, TX, USA, 17–20 July 2017; Available online: https://www.aiche.org/system/files/aiche-proceedings/conferences/404771/papers/485892/P485892.pdf (accessed on 9 January 2019).

19. Carrasco, F.; Grathwohl, S.; Maier, J.; Ruppert, J.; Scheffknecht, G. Experimental investigations of oxyfuel burner for cement production application. *Fuel* **2019**, *236*, 608–614. [CrossRef]

20. Jamali, A.; Fleiger, K.; Ruppert, J.; Hoenig, V.; Anantharaman, R. *Optimised Opearation of an Oxyfuel Cement Plant (D6.1)*. 2018. Available online: https://www.sintef.no/globalassets/project/cemcap/d-6.1-optimized-oxyfuel-operation.pdf.pdf (accessed on 8 January 2019).

21. Trædal, S.; Berstad, D. *Experimental Investigation of CO_2 Liquefaction for CO_2 Capture from Cement Plants (D11.2)*. 2018. Available online: https://www.sintef.no/globalassets/project/cemcap/2018-11-14-deliverables/d11.2-experimental-co2-liquefaction.pdf (accessed on 8 January 2019).

22. Pérez-Calvo, J.F.; Sutter, D.; Gazzani, M.; Mazzotti, M. Pilot tests and rate-based modelling of CO_2 capture in cement plants using an aqueous ammonia solution. *Chem. Eng. Trans.* **2018**, *69*, 145–150. [CrossRef]

23. Alonso, M.; Álvarez Criado, Y.; Fernández, J.R.; Abanades, C. CO_2 Carrying Capacities of Cement Raw Meals in Calcium Looping Systems. *Energy Fuels* **2017**, *31*, 13955–13962. [CrossRef]

24. Arias, B.; Alonso, M.; Abanades, C. CO_2 Capture by Calcium Looping at Relevant Conditions for Cement Plants: Experimental Testing in a 30 kWth Pilot Plant. *Ind. Eng. Chem. Res.* **2017**, *56*, 2634–2640. [CrossRef]

25. Turrado, S.; Arias, B.; Fernández, J.R.; Abanades, J.C. Carbonation of Fine CaO Particles in a Drop Tube Reactor. *Ind. Eng. Chem. Res.* **2018**, *57*, 13372–13380. [CrossRef]

26. Hornberger, M.; Spörl, R.; Scheffknecht, G. Calcium Looping for CO_2 Capture in Cement Plants—Pilot Scale Test. *Energy Procedia* **2017**, *114*, 6171–6174. [CrossRef]

27. Queneau, P.E.; Marcuson, S.W. Oxygen pyrometallurgy at copper cliff—A half century of progress. *J. Miner. Metals Mat. Soc.* **1996**, *48*, 14–21. [CrossRef]

28. IEAGHG. *Oxy Combustion Processes for CO_2 Capture from Power Plant*; Report Number 2005/9; IEAGHG: Cheltenham, UK, 2005; Available online: https://ieaghg.org/docs/General_Docs/Reports/Report%202005-9%20oxycombustion.pdf (accessed on 12 January 2019).

29. Locher, G. Mathematical models for the cement clinker burning process. Part 1–5. *Zement-Kalk-Gips* **2002**, *55*, 29–38.

30. Klein, H.; Hoenig, V. Model calculations of the fuel energy requirement for the clinker burning process. *Cem. Int.* **2006**, *3*, 44–63.

31. GECOS, GS Process Simulation Code. 2016. Available online: http://www.gecos.polimi.it/expertise/software-development/ (accessed on 12 January 2019).

32. Campanari, S.; Cinti, G.; Consonni, S.; Fleiger, K.; Gatti, M.; Hoppe, H.; Martínez, I.; Romano, M.; Spinelli, M.; Voldsund, M. *Design and Performance of CEMCAP Cement Plant without CO_2 Capture (D4.1)*; 2016. [CrossRef]

33. IEAGHG. *CO_2 Capture in the Cement Industry*; Report Number 2008/03; IEAGHG: Cheltenham, UK, 2008; Available online: https://ieaghg.org/docs/General_Docs/Reports/2008-3.pdf (accessed on 8 January 2019).

34. IEAGHG. *Evaluation of Reclaimer Sludge Disposal from Post-Combustion CO$_2$ Capture*; Report Number 2014/02; IEAGHG: Cheltenham, UK, 2014; Available online: https://ieaghg.org/docs/General_Docs/Reports/2014-02.pdf (accessed on 8 January 2019).

35. Knudsen, J.N.; Jensen, J.N.; Vilhelmsen, P.-J.; Biede, O. Experience with CO$_2$ capture from coal flue gas in pilot-scale: Testing of different amine solvents. *Energy Procedia* **2009**, *1*, 783–790. [CrossRef]

36. Koring, K. CO$_2$-Emissionsminderungspotential und Technologische Auswirkungen der Oxyfuel-Technologie im Zementklinkerbrennprozess. Ph.D. Thesis, Verein Deutscher Zementwerke (VDZ), Düsseldorf, Germany, 2013.

37. European Cement Research Academy GmbH (ECRA). *ECRA CCS Project—Report about Phase II*; Report Number TR-ECRA-106/2009; ECRA: Duesseldorf, Germany, 2009; Available online: https://www.ecra-online.org/fileadmin/redaktion/files/pdf/ECRA__Technical_Report_CCS_Phase_II.pdf (accessed on 8 January 2019).

38. European Cement Research Academy GmbH (ECRA). *ECRA CCS Project—Report on Phase III*; ECRA: Duesseldorf, Germany, 2012; Available online: https://ecra-online.org/fileadmin/redaktion/files/pdf/ECRA_39.Technical_Report_CCS_Phase_III.pdf (accessed on 8 January 2019).

39. Available online: Technical_Report_CCS_Phase_III.pdf (accessed on 8 January 2019).

40. Kemper, J.; Sutherland, L.; Watt, J.; Santos, S. Evaluation and Analysis of the Performance of Dehydration Units for CO$_2$ Capture. *Energy Procedia* **2014**, *63*, 7568–7584. [CrossRef]

41. Anantharaman, R. Energy Efficiency in Process Plants with Emphasis on Heat Exchanger Networks. Ph.D. Thesis, Norwegian University of Science and Technology, Trondheim, Norway, 2011.

42. Voldsund, M.; Anantharaman, R.; Berstad, D.; De Lena, E.; Fu, C.; Gardarsdottir, S.O.; Jamali, A.; Pérez-Calvo, J.F.; Romano, M.; Roussanaly, S.; et al. *CEMCAP Comparative Techno-Economic Analysis of CO$_2$ Capture in Cement Plants (D4.6)*. 2018. Available online: https://www.sintef.no/globalassets/project/cemcap/2018-11-14-deliverables/d4.6-cemcap-comparative-techno-economic-analysis-of-co2-capture-in-cement-plants.pdf (accessed on 8 January 2019).

43. Pérez-Calvo, J.F.; Sutter, D.; Gazzani, M.; Mazzotti, M. *Chilled Ammonia Process (CAP) Optimization and Comparison with Pilot Plant Tests (D10.3)*. 2018. Available online: https://www.sintef.no/globalassets/project/cemcap/2018-11-14-deliverables/d10.3_cap-optimization.pdf (accessed on 8 January 2019).

44. Berstad, D.; Trædal, S. *Membrane-Assisted CO$_2$ Liquefaction for CO$_2$ Capture from Cement Plants (D11.3)*. 2018. Available online: https://www.sintef.no/globalassets/project/cemcap/2018-11-14-deliverables/d11.3.pdf (accessed on 8 January 2019).

45. Romano, M.C. Modeling the carbonator of a Ca-looping process for CO$_2$ capture from power plant flue gas. *Chem. Eng. Sci.* **2012**, *69*, 257–269. [CrossRef]

46. Grasa, G.S.; Abanades, J.C.; Alonso, M.; González, B. Reactivity of highly cycled particles of CaO in a carbonation/calcination loop. *Chem. Eng. J.* **2008**, *137*, 561–567. [CrossRef]

47. Martínez, I.; Grasa, G.; Murillo, R.; Arias, B.; Abanades, J.C. Modelling the continuous calcination of CaCO3 in a Ca-looping system. *Chem. Eng. J.*. [CrossRef]

48. Spinelli, M.; De Lena, E.; Romano, M.C. *CaL Reactor Modelling and Process Simulations (D12.4)*; 2018. [CrossRef]

49. De Lena, E.; Spinelli, M.; Martínez, I.; Gatti, M.; Scaccabarozzi, R.; Cinti, G.; Romano, M.C. Process integration study of tail-end Ca-Looping process for CO$_2$ capture in cement plants. *Int. J. Greenh. Gas Control* **2017**, *67*, 71–92. [CrossRef]

50. Spinelli, M.; Martínez, I.; Romano, M.C. One-dimensional model of entrained-flow carbonator for CO$_2$ capture in cement kilns by Calcium looping process. *Chem. Eng. Sci.* **2018**, *191*, 100–114. [CrossRef]

51. Grasa, G.S.; Murillo, R.; Alonso, M.; Abanades, J.C. Application of the random pore model to the carbonation cyclic reaction. *AIChE J.* **2009**, *55*, 1246–1255. [CrossRef]

52. De Lena, E.; Spinelli, M.; Gatti, M.; Scaccabarozzi, R.; Campanari, S.; Consonni, S.; Cinti, G.; Romano, M.C. Techno-economic analysis of calcium looping processes for low CO$_2$ emission cement plants. *Int. J. Greenh. Gas Control* **2019**, *82*, 244–260. [CrossRef]

53. Van Der Spek, M.; Roussanaly, S.; Rubin, E. Best practices and recent advances in CCS cost engineering and economic analysis. *Int. J. Greenh. Gas Control* **2019**. [CrossRef]

54. Knudsen, J.N. Results and Future Perspectives of Aker Solutions' Amine Project. In Proceedings of the Norcem International CCS Conference, Langesund, Norway, 20–21 May 2015.

55. Carrasco, F.M.; Grathwohl, S.; Maier, J.; Wilms, E.; Ruppert, J. *Oxyfuel Burner Prototype Performance Tests (D7.2)*. 2018. Available online: https://www.sintef.no/globalassets/project/cemcap/presentasjoner/d7.2-burner-tests_revision1.pdf (accessed on 12 January 2019).
56. Paneru, M.; Mack, A.; Maier, J.; Cinti, G.; Ruppert, J. *Oxyfuel Suspension Calciner Test Results (D8.2)*. 2018. Available online: https://www.sintef.no/globalassets/project/cemcap/cemcap-d-8-2-ustutt_final.pdf (accessed on 12 January 2019).
57. Lindemann Lino, M.; Matthias, B.; Ruppert, J.; Hoenig, V.; Becker, S.; Mathai, R. Analysis of Oxyfuel Clinker Cooler Operational Performance (D9.2). 2018. Available online: https://www.sintef.no/globalassets/project/cemcap/presentasjoner/d9.2_revision1_final.pdf (accessed on 12 January 2019).
58. Telikapalli, V.; Kozak, F.; Francois, J.; Sherrick, B.; Black, J.; Muraskin, D.; Cage, M.; Hammond, M.; Spitznogle, G. CCS with the Alstom chilled ammonia process development program–Field pilot results. *Energy Procedia* **2011**, *4*, 273–281. [CrossRef]
59. Baburao, B.; Kniesburger, P.; Lombardo, G. Chilled Ammonia Process Operation and Results from Pilot Plant at Technology Centre Mongstad, In Proceedings of the Trondheim CCS Conference (TCCS-8), Trondheim, Norway, 16–18 June 2015.
60. Hägg, M.-B.; Lindbråthen, A.; He, X.; Nodeland, S.G.; Cantero, T. Pilot demonstration—Reporting on CO_2 capture from a cement plant using hollow fiber process, In Proceedings of the 13th International Conference on Greenhouse Gas Control Technologies, Lausanne, Switzerland, 14–18 November 2016.
61. Arias, B.; Diego, M.E.; Abanades, J.C.; Lorenzo, M.; Diaz, L.; Martínez, D.; Alvarez, J.; Sánchez-Biezma, A. Demonstration of steady state CO_2 capture in a 1.7MWth calcium looping pilot. *Int. J. Greenh. Gas Control* **2013**, *18*, 237–245. [CrossRef]

 *energies*

Article

Comparison of Technologies for CO$_2$ Capture from Cement Production—Part 2: Cost Analysis

Stefania Osk Gardarsdottir [1,*]**, Edoardo De Lena , Matteo Romano** [2]**, Simon Roussanaly** [1]**,
Mari Voldsund** [1]**, José-Francisco Pérez-Calvo** [3]**, David Berstad** [1]**, Chao Fu** [1]**,
Rahul Anantharaman** [1]**, Daniel Sutter** [3]**, Matteo Gazzani** [4]**, Marco Mazzotti** [3] **and Giovanni Cinti** [5]

[1] SINTEF Energy Research, Department of Gas Technology, NO-7465 Trondheim, Norway;
    simon.roussanaly@sintef.no (S.R.); mari.voldsund@sintef.nomailto (M.V.); david.berstad@sintef.no (D.B.);
    chao.fu@sintef.no (C.F.); rahul.anantharaman@sintef.no (R.A.)

[2] Politecnico di Milano, Department of Energy, 20156 Milan, Italy; edoardo.delena@polimi.it (E.D.L.);
    matteo.romano@polimi.it (M.R.)

[3] Institute of Process Engineering, Department of Mechanical and Process Engineering, ETH Zurich,
    8092 Zurich, Switzerland; francisco.perezcalvo@ipe.mavt.ethz.ch (J.-F.P.-C.); sutter@ipe.mavt.ethz.ch (D.S.);
    marco.mazzotti@ipe.mavt.ethz.ch (M.M.)

[4] Copernicus Institute of Sustainable Development, Energy and Resources, Utrecht University,
    3584 CB Utrecht, The Netherlands; m.gazzani@uu.nl

[5] Italcementi Heidelberg Group, 24126 Bergamo, Italy; g.cinti@italcementi.it

* Correspondence: stefania.gardarsdottir@sintef.no; Tel.: +47-412-22-758

Received: 5 December 2018; Accepted: 30 January 2019; Published: 10 February 2019

Abstract: This paper presents an assessment of the cost performance of CO$_2$ capture technologies when retrofitted to a cement plant: MEA-based absorption, oxyfuel, chilled ammonia-based absorption (Chilled Ammonia Process), membrane-assisted CO$_2$ liquefaction, and calcium looping. While the technical basis for this study is presented in Part 1 of this paper series, this work presents a comprehensive techno-economic analysis of these CO$_2$ capture technologies based on a capital and operating costs evaluation for retrofit in a cement plant. The cost of the cement plant product, clinker, is shown to increase with 49 to 92% compared to the cost of clinker without capture. The cost of CO$_2$ avoided is between 42 €/t$_{CO2}$ (for the oxyfuel-based capture process) and 84 €/t$_{CO2}$ (for the membrane-based assisted liquefaction capture process), while the reference MEA-based absorption capture technology has a cost of 80 €/t$_{CO2}$. Notably, the cost figures depend strongly on factors such as steam source, electricity mix, electricity price, fuel price and plant-specific characteristics. Hence, this confirms the conclusion of the technical evaluation in Part 1 that for final selection of CO$_2$ capture technology at a specific plant, a plant-specific techno-economic evaluation should be performed, also considering more practical considerations.

Keywords: CCS; cement; techno-economic analysis; MEA-based absorption; chilled ammonia; membrane-assisted CO$_2$ liquefaction; oxyfuel; calcium looping

---

1. Introduction

Production of cement is estimated to account for about 7% of anthropogenic CO$_2$ emissions, thus contributing significantly to climate change [1]. Approximately 2/3rd of the CO$_2$ emissions are process related, originating from the conversion of limestone, CaCO$_3$ to CaO and CO$_2$, while the remaining 1/3rd comes from the combustion of fuels in the rotary kiln of the cement plant. A recent technology roadmap published by the International Energy Agency and the Cement Sustainability Initiative, a global consortium of 24 major cement producers, identified several main carbon mitigation options for the cement industry [1]. These include e.g., reduction of clinker to cement ratio, fuel switching and implementation of CO_2 capture and storage (CCS). Implementation of CCS was found to have the largest CO_2 emission reduction potential of the mitigation options due to its ability to drastically reduce both process and fuel related emissions from cement kilns. Combining CCS with CO_2 utilization (CCUS) is also being discussed as an alternative emission mitigation option and a business case although recent studies suggest that less than 10% of the captured CO_2 in a cement plant could be economically converted to added-value products [2].

Although cement plants are moderately large emission sources compared to large-scale fossil-fueled power plants, they possess several characteristics favorable for CO_2 capture, such a relatively high CO_2 concentration in their flue gases, few emission points, stable operation and, in some specific cases, available waste heat. The cement industry has been showing increased interest in CO_2 capture technologies in recent years, especially in Europe where the European Cement Research Academy (ECRA) has actively carried out CCS research since 2007 [3]. CCS applied to cement production has gained further interest after testing at the Norcem Brevik plant in Norway, which has been selected as one of the two potential sites for CO_2 capture in the in the Norwegian full-scale CCS project. On-site pilot testing included three CO_2 capture technologies: amine absorption, amine-impregnated adsorption and fixed-site carrier membranes [4–7]. Presently, a front-end engineering design (FEED)-study for the Norcem Brevik plant is being carried out to prepare for a final investment decision by the Norwegian Parliament in 2020/2021 [8].

The increasing interest in CCS from the cement industry has resulted in the publication of several studies investigating techno-economic performance of different CO_2 capture technologies integrated in cement plants. Most of the studies have focused on retrofitting amine-based CO_2 capture processes [9–15], while few studies have also considered a case of new construction [10,11]. The supply of heat to the capture process also varies between the studies, e.g., Liang and Li [9] considered investment in a small coal-fired combined heat and power (CHP) plant for steam and electricity supply while IEAGHG [10] considered steam generation from waste heat together with supply from either a natural gas boiler or a CHP plant with export of surplus electricity. Furthermore, in their plant-specific study, Jakobsen et al. [13] considered cases where only available waste heat was used for partial-scale capture as well as a case with waste heat and additional steam production from a natural gas boiler for full-scale capture.

Calcium-based looping systems have been studied in several configurations with different strategies for waste heat utilization. Studies on indirect calcination configuration with a relatively low CO_2 avoidance rate have either considered all electricity to be imported (Ozcan [11]) or included investments in waste heat recovery systems for electricity generation (Rodríguez et al. [16] and Diego et al. [17]). Calcium looping in tail-end like configurations with waste heat steam generation and high CO_2 avoidance rate have also been investigated by Ozcan [11] and Rodríguez et al. [16], and yet another configuration, double calcium looping with waste heat recovery, was proposed by Diego et al. [17].

Oxyfuel combustion with CO_2 capture was investigated by the IEAGHG [10] in partial and full capture configurations, and more recently by Gerbelová, van der Spek and Schakel [14] for full capture. In both studies, electricity was imported from the grid. Both studies also highlighted the potential for significant cost-reduction with oxyfuel compared with MEA-based amine capture but emphasized the large modifications required in the core cement process for implementing the oxyfuel technology and the uncertain impacts on product quality.

A few studies have also investigated the techno-economics of membrane-based technologies for application in cement plants. Lindqvist et al. [18] investigated multi-stage dense polymeric membrane and facilitated transport membrane, Ozcan [11] evaluated a dual-stage polymeric membrane and Jakobsen et al. [13] investigated multi-stage polymeric membrane and a fixed site carrier membrane in their plant-specific study. These studies highlighted the potential for a relatively low-cost membrane system compared with MEA-based amine capture. However, Jakobsen, et al. emphasized the need for further technology demonstration to reduce uncertainties.

Overall, a wide range in CO_2 avoidance costs was observed for the different capture technologies. The level of detail of the techno-economic evaluations, the methodologies and the assumptions used varies considerably. It is therefore difficult to make a direct comparison of techno-economic performance of different CO_2 capture technologies applied in cement plants from literature sources in order to identify the best CO_2 capture options.

In this paper, the economic performance of CO_2 capture technologies retrofitted in a Best-Available-Technologies (BAT) cement plant are assessed in the context of a coherent techno-economic framework [19]. The investigated capture technologies are MEA-based absorption as reference technology, chilled ammonia process (CAP), membrane-assisted CO_2 liquefaction, oxyfuel technology and two different configurations of calcium looping technology (tail-end and integrated). Besides highlighting the value of the consistent application of the above-mentioned common framework in a comparative investigation of a broad set of CO_2 capture technologies, it is worth noting that for several of these technologies this work represents the first detailed costs analysis for a cement application.

This work has been carried out within the Horizon2020-funded project CEMCAP, which has as overall objective to prepare the ground for large-scale implementation of CO_2 capture in the European cement industry [20]. An essential element in responding to this objective has been to perform a comprehensive techno-economic comparative assessment of CO_2 capture. With this paper we aim at providing an assessment that can be used as a decision basis for future evaluations of CO_2 capture implementation at cement plants. An extraction of this work is presented as this paper series, where the technical evaluation is in Part 1, and the economic analysis is in Part 2.

2. Reference Cement Plant and CO_2 Capture Technologies

The reference cement plant is a BAT plant defined by ECRA. It is based on a dry kiln process, and consists of a five-stage cyclone preheater, calciner with tertiary duct, rotary kiln and clinker cooler. Flue gas is emitted from a single stack with CO_2 emissions originating from combustion of fuel in the calciner and the rotary kiln, as well as from the calcination of the raw material itself ($CaCO_3 \rightarrow CaO + CO_2$). Some waste heat can be recovered in the cement plant from the clinker cooler exhaust air. Compared to ECRA reference, here a selective non-catalytic reduction (SNCR) system for deNOx removal is considered.

The plant has a representative size for a European cement plant with a capacity of 3000 tons of clinker per day. This corresponds to a capacity of about 1 Mt clinker per year, with a run time of >330 days per year. The specific CO_2 emissions and the electric power consumption of the plant amount to 850 $kgCO_2/t_{clk}$ (18–22 vol% CO_2 in flue gas, on wet basis) and 132 kWh/t_{clk}, respectively. The clinker burning line of the reference cement plant is shown in Figure 1. A more detailed description of the reference cement plant can be found in Part 1 of this study and in the CEMCAP framework [19]. The utility and material consumption of the reference cement plant (based on the process modelling presented in Part 1) are summarized in Table 1.

Table 1. Utilities and consumables for the reference cement plant without CO_2 capture.

Utility and Consumable	Value
Clinker production (t/h)	120.65
Coal (t/h)	13.93
Electric power (MW)	15.88
Ammonia solution for NO_x reduction (t/h)	0.60

The investigated CO_2 capture technologies are fundamentally different, both in terms of the capture concepts themselves, but also when it comes to the inputs required (coal, heat, electric power), whether electric power is consumed or generated, and the way the capture technologies are integrated into the kiln (ranging from purely end-of-pipe to considerable modification of the process at the cement kiln). A schematic overview of the integration of the capture technologies to the reference kiln is given in Figure 2, followed by a brief description of each technology. More detailed descriptions of the technologies can be found in Part 1 of the paper series.

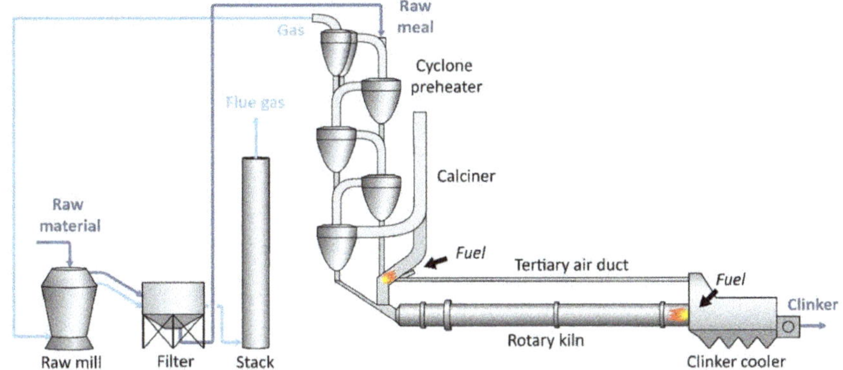

Figure 1. The clinker burning line of the CEMCAP reference cement plant.

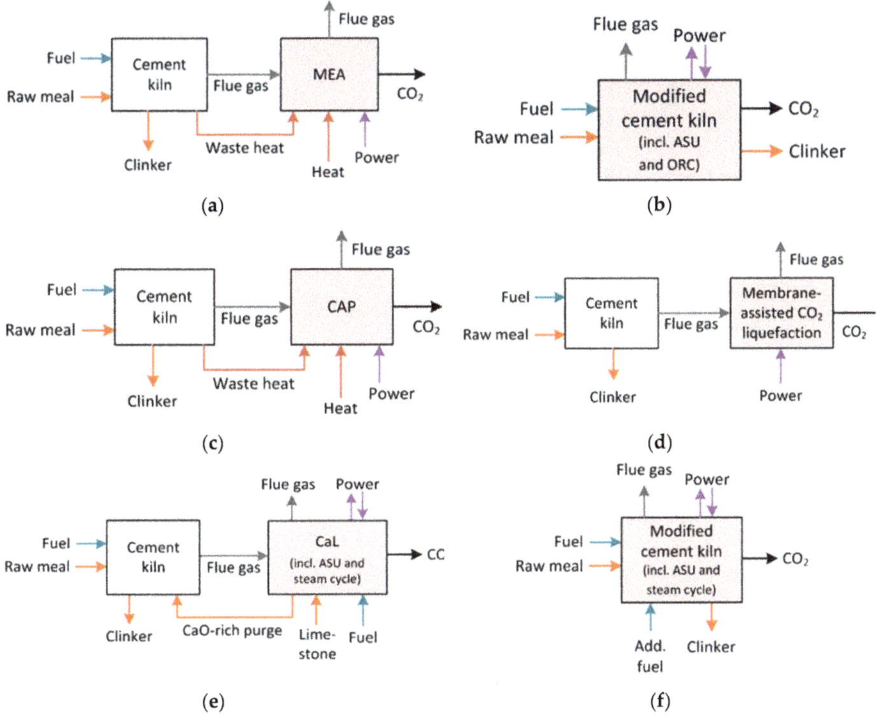

Figure 2. Schematic overview over investigated technologies (pink) and their integration into the reference kiln (white). (**a**) Reference technology: MEA; (**b**) Oxyfuel; (**c**) Chilled ammonia process; (**d**) Membrane-assisted CO_2 liquefaction; (**e**) Calcium looping-tail-end; (**f**) Calcium looping integrated.

The reference technology MEA is an end-of-pipe technology based on absorption. The MEA process requires a considerable amount of heat for solvent regeneration, and power is required for fans and pumps in the core process as well as for compression and dehydration of the captured CO_2.

The flue gas from the cement plant is treated in the capture system right before it reaches the stack, and waste heat from the cement plant is used to cover a part of the heat demand.

In the oxyfuel process, combustion is performed with an oxidizer consisting mainly of oxygen mixed with recycled CO_2, to produce a CO_2 rich flue gas which allows a relatively easy purification with a CO_2 purification unit (CPU). As opposed to the MEA technology, the cement kiln process is modified when the oxyfuel process is integrated into a kiln system. Additional power is needed for an air separation unit (ASU) and for the CPU, but some of this power demand can be covered by an organic Rankine cycle (ORC) generating power from waste heat.

The chilled ammonia process (CAP) is also an end-of-pipe technology based on absorption, where CO_2 is removed from flue gas using aqueous ammonia as solvent. Heat is required for solvent regeneration and for an ammonia recovery system and power is required for chilling, pumping and compression. Waste heat from the cement plant can be utilized to cover a part of the heat demand.

In the membrane-assisted CO_2 liquefaction (MAL) concept, polymeric membrane technology and a CO_2 liquefaction process are combined since CO_2 liquefaction is generally more suitable than membranes for second-stage CO_2 purification [21]. Polymeric membranes are first utilized for bulk separation of CO_2 resulting in moderate product purity. This CO_2-rich product is sent to the CO_2 liquefaction process, where CO_2 is liquefied, and the more volatile impurity components are removed, resulting in a high purity CO_2 product. The technology is an end-of-pipe retrofit technology with no additional integration or feedback to the cement plant, and only electric power is required as input to the process.

The calcium looping (CaL) technology is based on the reversible carbonation reaction, which is exploited to separate the carbon dioxide from the flue gas. The technology can be applied to a cement plant as a tail-end/end-of-pipe technology (CaL tail-end, Figure 2e) or it can be integrated with the calcination process taking place in the cement kiln (CaL integrated, Figure 2f). In the tail-end configuration the flue gas from the cement kiln is sent to the CaL system for purification, and a CaO-rich purge from the CaL system is sent to the cement kiln and added to the raw meal. In the integrated concept, the calciner and the preheater of the cement kiln are modified (the cement kiln calciner and the CaL calciner are combined), and the CO_2 capture is performed as a part of the process. The CaL processes require supply of limestone and coal. Oxygen is required for oxyfuel combustion in the calciner. Power is required for an ASU for oxygen supply to the core CaL process and for a CPU. A steam cycle recovers high temperature waste heat and produces power that can cover demand of the process and/or be exported.

The cost analysis of the CO_2 capture retrofit considers 90% CO_2 captured from the flue gas at the stack in the reference cement plant as a baseline scenario. Furthermore, the captured CO_2 is compressed and conditioned for transport by pipeline. The required CO_2 pressure is 110 bar and the temperature is around 30 °C. Further details on requirements for CO_2 purity and maximum impurity concentrations are outlined in the CEMCAP framework [19]. For the capture technologies that require steam in their operation, available waste heat from the cement plant is used to cover as much of the steam demand as possible while the rest of the steam required (the major part) is generated by a natural gas boiler (see Part 1 for details).

Utility and material consumption of the CO_2 capture technologies as well as equivalent specific CO_2 avoided for all technologies, based on process simulations presented in Part 1, are summarized in Table 2. It should be mentioned that the oxyfuel and CaL technologies are closely integrated with the cement kiln, while the other technologies are only connected to the kiln by the flue gas entering the system and heat integration. Due to the close process integration, the two CEMCAP partners simulating the oxyfuel and CaL technologies, VDZ and PoliMi, have established their own simulations of the reference cement kiln. Other technologies are simulated using the flue gas from the VDZ simulation of the reference kiln as feed.

Table 2. Utilities, consumables and CO_2 avoided for the cement plant with CO_2 capture.

	MEA	Oxyfuel	CAP	MAL	CaL-Tail-End	CaL-Integrated
Clinker production (t/h)	120.7	125.0	120.7	120.7	117.7	117.4
Coal (t/h)	13.9	14.5	13.9	13.9	30.8	23.5
Electric power (MW)	29.5	35.1	24.2	50.0	6.8	20.4
Steam from waste heat (MW)	3.7	-	4.7	-	-	-
Steam from NG boiler (MW)	92.7	-	56.1	-	-	-
Cooling water make-up (t/h)	208.2	104.5	185.7	85.3	256.3	263.5
MEA make-up (t/h)	0.1	-	-	-	-	-
Process water make-up (t/h)	46.0	-	1.1	-	-	-
NaOH solution for DeSOx (t/h)	0.1	-	-	0.1	-	-
Ammonia solution for SNCR (t/h)	0.6	0.6	0.6	0.6	0.6	0.6
Ammonia solvent make-up (t/h)	-	-	0.2	-	-	-
Sulfuric acid for ammonia recovery (t/h)	-	-	0.1	-	-	-
Membrane material replacement (m^2/year)	-	-	-	50,160	-	-
Equivalent specific CO_2 avoided (kgCO_2/t$_{clk}$)	559	719	640	687	806	797

3. Methodology

The economic assessment of retrofitting CO_2 capture technologies in a BAT cement plant is based on the results from the detailed technical process evaluations for each of the technologies described in Part 1. The technical process evaluations are based on process simulations with input from experimental work carried out in the CEMCAP project, on the oxyfuel technology [22,23], membrane-assisted CO_2 liquefaction [24], the chilled ammonia process [25,26] and calcium looping [27–30]. Economic key performance indicators (KPIs) are finally calculated and used to compare the techno-economic performance of the technologies.

3.1. Cost Estimation

The cost estimation is performed on the basis of earnings before interest, taxes, depreciation and amortization. The estimation consists of two main parts: estimation of (i) the capital costs (CAPEX) which is expressed in terms of total plant cost (TPC), and (ii) the operating costs (OPEX). All cost figures are expressed in €$_{2014}$. The main assumptions and descriptions of cost elements that make up the CAPEX and OPEX are summarized in this section. For further details the reader is referred to the CEMCAP framework [19]. Furthermore, a spreadsheet with the model developed in this work for the cost estimation is available for open use [31].

3.1.1. Capital Costs

A bottom-up approach is used for estimation of TPC for the CO_2 capture technologies [32]. The cost estimates of all the CO_2 capture technologies are performed for "Nth of a kind" plants, i.e., for commercial plants built after successful development and commercial adoption of the technology. A breakdown of the costing approach is illustrated in Figure 3.

Estimation of total equipment costs (TEC) and installation costs are based on equipment lists compiled for each of the investigated CO_2 capture technologies (see Supplementary Material for detailed equipment lists). Estimation of equipment costs (EC) and installation costs (IC) for most standard process equipment is done using Aspen Process Economic Analyzer® and the Thermoflex® software. The estimation is based on key characteristics of each equipment from process simulations and design criteria, such as pressure, temperature, flows and materials. Estimates for other, non-standard components are based on information provided by the CEMCAP industry partners and literature. This includes e.g., several pieces of non-commercial process equipment in the oxyfuel and CaL systems, membrane packages and multi-stream plate and fin heat exchangers used in CO_2 purification units (CPU). More details on design criteria for standard process equipment and cost estimation methodologies for non-standard equipment can be found in CEMCAP report D4.4 "Cost of critical components in CO_2 capture processes" [33].

Estimation of TEC and installation costs for the CAP technology is performed by the project partner Baker Hughes, a GE company, (BHGE, Frankfurt am Main, Germany) using their proprietary tool QFACT which is based on an extensive database of executed projects. The unit costs are lumped into equipment costs and installation costs as to not disclose BHGE confidential information about cost structure and/or pricing strategy.

Process contingencies are based on the maturity or status of the technology, in line with the American Association of Cost Engineers (AACE) guidelines for process contingency [34] and are adjusted to also account for the estimated level of detail of the equipment lists for each technology. The resulting process contingency factors for each of the capture technologies and miscellaneous subsystems are listed in Table 3.

Table 3. Process contingency factors for core CO_2 capture technologies and miscellaneous subsystems.

Technology	Process Contingency—Maturity (% of TDC')	Process Contingency—Detail Level of Equipment List (% of TDC')
MEA	15	3
Oxyfuel	30	12
CAP	20	0
MAL	40	12
CaL tail-end	20	12
CaL integrated	60	12
ASU	5	0
Cooling systems	5	0
Refrigeration systems	5	Same as CO_2 capture technology
CO_2 purification units	20	Same as CO_2 capture technology

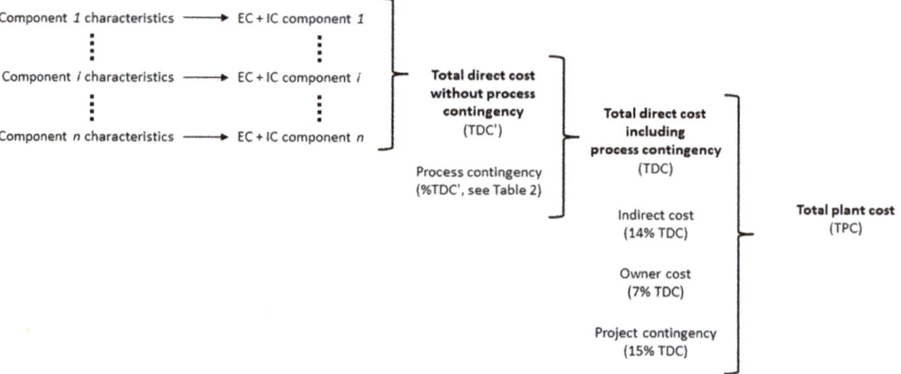

Figure 3. Break-down of cost elements in the bottom-up approach for estimation of total plant costs [35].

Indirect costs are set to 14% of the total direct costs (TDC) for all technologies and include cost elements such as yard improvement, service facilities, engineering and consultancy cost as well as building and sundries [32].

Owner's costs and project contingencies for Nth of a kind cost estimates are set to 7% and 15% of the TDC, respectively, following the AACE cost estimates guidelines.

The accuracy of the cost estimate is expected to be +35%/−15% (AACE Class 4), except for the CAP technology, where the estimation of TEC and installation cost is performed by BHGE with expected accuracy of ±30% (also AACE Class 4).

3.1.2. Operating Costs

Fixed OPEX, which include maintenance, insurance and labour costs are based on assumptions for material replacement and factor approach [32]. The annual maintenance cost is taken as 2.5% of the TPC and includes cost of preventive and corrective maintenance as well as maintenance labour cost. Maintenance labour cost corresponds to 40% of the total annual maintenance cost. The annual insurance and location taxes, including overhead and miscellaneous regulatory fees are set to 2% of TPC. Labour costs include costs for operating, administrative and support labour. Costs for operating labour are calculated from assumptions on number of employees, 100 persons in the cement plant and 20 persons in the CO_2 capture plant, with an annual fully-burdened cost per employee of 60 k€/person. Costs for administrative and support labour are assumed to be 30% of the operating and maintenance labour cost.

Variable OPEX, which include fuel and raw material costs, utilities and other consumables, are primarily based on process simulations. No carbon tax is considered in the calculation of variable OPEX. The unit cost of all materials and utilities considered in the cost analysis are listed in Table 4.

Table 4. Unit cost of materials and utilities used in the cost analysis.

Variable OPEX Item	Unit Cost
Raw meal price (€/t_{clk})	5
Coal price (€/GJ_{LHV})	3
Natural gas price (€/GJ_{LHV})	6
Price of electricity (€/MWh)	58.1
Cost of the steam produced from a natural gas boiler (€/MWh)	25.3
Cost of the steam produced from the cement plant waste heat (€/MWh)	8.5
Cooling water cost (€/m^3)	0.39
Process water cost (€/m^3)	6.65
Ammonia solution price for NO_x removal (€/t)	130
MEA solvent (€/t)	1450
Ammonia solvent (€/t)	406
Sulfuric acid (€/t)	46
Sodium hydroxide for flue gas desulfurization (€/t)	370
Membrane material replacement (€/m^2)	7.87
Miscellaneous variable O&M (€/t_{clk})	1.1

3.2. Economic Key Performance Indicators

The cost performance of the capture technology is evaluated by the cost of clinker and the cost of CO_2 avoided. In calculating the KPIs, the economic boundaries and financial parameters listed in Table 5 are used.

Table 5. Economic boundaries and financial parameters used in calculating economic KPIs.

Capacity factor (%)	91.3
Economic life (years)	25
Construction time, cement plant (years)	2
Allocation of cement plant construction costs by year (%)	50/50
Construction time—CO_2 capture (years)	3
Allocation of CO_2 capture construction costs by year [1] (%)	40/30/30
Discount rate (%)	8

[1] For certain CO_2 capture technologies, like the oxyfuel and integrated CaL technologies, a significant downtime might be required to modify the existing cement plant for deep integration with the CO_2 capture plant. Although this could impact the cost performance of these technologies, this is not considered here due to the lack of publicly available knowledge and highly site-specific nature of this issue.

The cost of clinker (*COC*) is evaluated by summing the contributions of the annualized CAPEX C_{cap}, of the fuel cost C_{fuel}, of the raw material costs C_{RM}, of the electricity cost C_{el}, and of the other operating and maintenance cost $C_{O\&M}$, all expressed per ton of clinker produced (i.e., as $€/t_{clk}$). In case the cement plant has a net power export, revenues for electricity export to the grid are considered and C_{el} becomes negative:

$$COC = C_{cap} + C_{fuel} + C_{RM} + C_{el} + C_{O\&M} \qquad (1)$$

The cost of CO_2 avoided (CAC), in $€/t_{CO2}$, is evaluated based on the cost of clinker and the equivalent specific emissions of the cement plant with and without CO_2 capture as shown in Equation (2) [36],

$$CAC = \frac{COC - COC_{ref}}{e_{clk,eq,ref} - e_{clk,eq}} \qquad (2)$$

where $e_{clk,eq,ref}$ is specific equivalent emissions from the reference cement plant, in t_{CO2}/t_{clk}, and $e_{clk,eq}$ is the specific equivalent emission from the cement plant with capture.

Equivalent emissions are defined as the sum of direct e_{clk} and indirect $e_{el,clk}$ emissions:

$$e_{clk,eq} = e_{clk} + e_{el,clk} \qquad (3)$$

Indirect emissions can be calculated using the following equation:

$$e_{el,clk} = e_{el} \cdot P_{el,clk} \qquad (4)$$

where $P_{el,clk}$ is the specific power consumption, which is positive when power is consumed and negative when it is generated, and e_{el} is the CO_2 emissions associated with each unit of electric power consumed. This value depends largely on the electricity mix considered.

The equivalent CO_2 avoided takes all direct and indirect emissions into account. It gives the best indication on the overall reduction in CO_2 emissions of the cement plant when a certain capture technology is implemented and allows a fair comparison of different technologies.

3.3. Economic Data of the Reference Cement Plant

The TDC of the reference cement plant is based on estimations from the IEAGHG [10] for a BAT cement plant with the same clinker capacity as the CEMCAP reference plant and amounts to 149.8 M€$_{2014}$. This includes the added costs of a DeNOx system, based on standard SNCR process, assumed to be installed in the reference cement plant. The SNCR system uses ammonia solution as a reduction agent and has an average reduction rate of 60%. The TDC for the SNCR system is assumed to be 1.01 M€$_{2014}$. The TPC for the reference cement plant are consequently calculated according to the bottom-up approach described in Figure 3.

4. Results and Discussion

4.1. Comparative Analysis of Key Performance Indicators

The KPIs employed to evaluate the economic performance of the cement plant with CO_2 capture are the cost of clinker and the cost of CO_2 avoided. The economic KPIs, as well as the total plant costs and annual OPEX for all the capture technologies and the reference cement plant without CO_2 capture are presented in Table 6. Detailed equipment lists with estimated equipment costs and direct costs on component basis are provided as Supplementary Material.

The cost evaluation shows that the reference capture technology, MEA, has the lowest total plant cost but also the highest annual OPEX. The MAL technology has the highest total plant cost, roughly three times higher the MEA technology. The oxyfuel and both CaL technologies have relatively low OPEX. In general, the cost of clinker increases with 49–92% from the 62.6 $€/t_{clk}$ in the reference cement plant when the investigated CO_2 capture technologies are implemented. The cost of CO_2 avoided ranges from 42 $€/t_{CO2}$ for the oxyfuel technology to 84 $€/t_{CO2}$ for the MAL technology, which is on a similar level as the CO_2 avoidance cost for the MEA reference technology.

Table 6. Summary of total plant costs and economic KPIs for the reference cement plant and the CO_2 capture technologies.

	Ref. Cement Plant	MEA	Oxyfuel	CAP	MAL	CaL-Tail-End	CaL-Integrated
TPC, cement plant + CO_2 capture plant (M€)	204	280	332	353	450	406	424
TPC, CO_2 capture plant (M€)	-	76	128	149	247	202	220
Annual OPEX (M€)	41	76	58	66	71	59	61
Cost of clinker (€/t_{clk})	62.6	107.4	93.0	104.9	120.0	105.8	110.3
Cost of CO_2 avoided (€/t_{CO2})	N/A	80.2	42.4	66.2	83.5	52.4	58.6

Figures 4 and 5 show the breakdown of the cost of clinker and of CO_2 avoided into the main cost factors. The oxyfuel technology shows the lowest cost of clinker compared to the other CO_2 capture technologies, both due to lower variable OPEX and lower CAPEX. The absorption-based technologies MEA and CAP as well as both CaL technologies have similar clinker costs, in the range of 105–110 €/t_{clk}. The CaL tail-end technology produces a significant amount of electricity which covers the electricity demand of the CO_2 capture process as well as a part of the cement plant's demand. As a result, this technology shows a lower electricity cost per ton clinker than the reference cement plant. The MAL technology shows the highest cost of clinker, with CAPEX and fixed OPEX (directly related to the CAPEX) being the largest cost factors.

The most important contributions to the cost of CO_2 avoided differ among the capture technologies and illustrate the fundamental differences between most of the technologies. In the case of the reference technology MEA, steam contributes most to the cost of CO_2 avoided. The consumption of steam is responsible for a large increase in the cost of clinker compared to the reference cement plant. Additionally, it has a negative effect on the equivalent specific CO_2 avoided due to the emissions from the natural gas boiler. For practical reasons, the boiler flue gas is not treated in the capture plant according to the common framework, as mixing it with the cement flue gas has detrimental effects on the CO_2 concentration. Compared with the other capture technologies, the MEA-based capture has the lowest equivalent specific CO_2 avoided, as can be seen in Table 2.

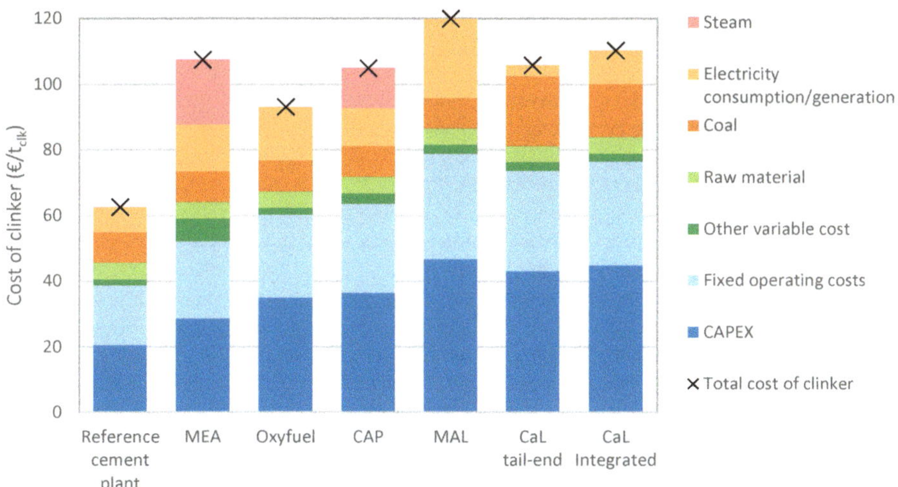

Figure 4. Break-down of cost of clinker for the reference cement plant and all the investigated CO_2 capture technologies.

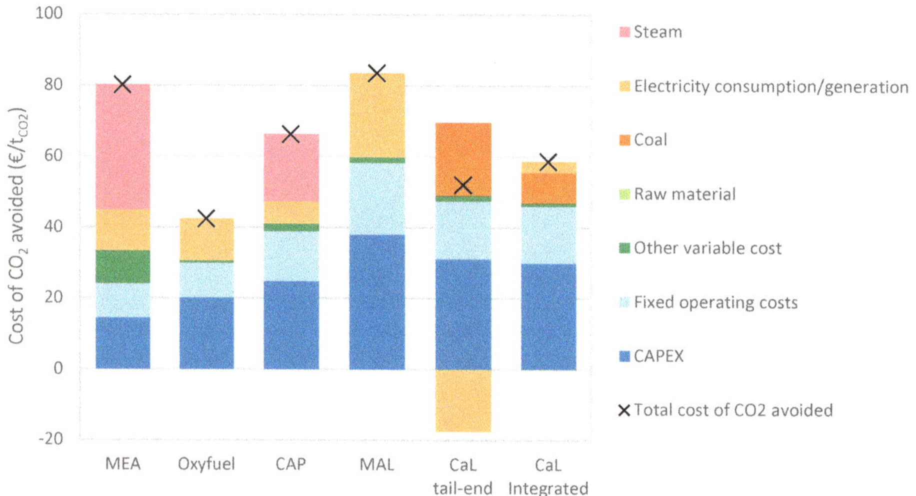

Figure 5. Break-down of cost of CO_2 avoided for all the investigated CO_2 capture technologies.

The oxyfuel technology has by far the lowest cost of CO_2 avoided. It is mainly the CAPEX, and associated fixed OPEX, together with electricity consumption in the capture process that contribute to the CO_2 avoided cost. The increase in electricity consumption contributes not only to an increase in the cost of clinker compared to the reference cement plant but also to a decrease in the equivalent specific CO_2 avoided due to associated CO_2 emissions.

In the case of the CAP, the cost of steam, as well as the CAPEX and fixed OPEX, are the most important factors. Compared to MEA, the cost of steam is significantly lower for the CAP due to its relatively low specific heat requirement. Hence, the equivalent specific CO_2 avoided of the CAP technology is about 15% higher than for MEA (cf. Table 2). This contributes to the lower CO_2 avoidance cost observed for the CAP technology compared to the MEA reference technology.

For MAL, high CAPEX and associated fixed OPEX contribute the most to the cost of CO_2 avoided. A significant share of the cost can also be attributed to the considerable electricity consumption of the capture process and the associated indirect CO_2 emissions which consequently have a negative effect on the equivalent specific CO_2 avoided.

For both calcium looping technologies, the increase in coal consumption compared with the reference cement plant contributes significantly to the cost of CO_2 avoided, together with the increase in CAPEX. Both CaL technologies generate a significant amount of electric power, with the generation in the tail-end case even covering the demand of the capture process and a part of the cement plant's demand. As a result, the cost of electricity per ton clinker is lower in the CaL tail-end case compared with the reference cement plant. This in turn leads to negative CO_2 avoidance costs associated with electricity consumption, as shown in Figure 5. For an extensive discussion on the economic analysis of the CaL cement plants, the reader is referred to the study of De Lena et al. [37].

Several studies on economic assessments of CO_2 capture from cement have been published in the literature and recently gathered in a review by the IEAGHG [38]. The results presented here are in line with the literature, although a direct comparison of cost estimates is challenging, due to variations in the level of detail, in the methodology and in the assumptions applied by the different studies. Most cost analyses have been carried out for MEA-based CO_2 capture, where various process configurations and assumptions have been considered. Therefore, a large range of CO_2 avoidance cost is reported for this technology, from around 75–170 €/t_{CO2} [38].

Fewer studies have analysed the cost of the oxyfuel technology. The IEAGHG [10] reported lower CAPEX than in this work (around 14% lower in €/t_{clk}), with the main difference being a lower cost estimated for the ASU and the CPU. On the other hand, the study reported around 8% higher CO_2 avoidance cost than in this work. Gerbelová, van der Spek and Schakel [14] estimated the CAPEX to be around 2% lower than estimated in this work, but they did not report on the CO_2 avoidance cost.

Ozcan [11] reported on CAPEX for a calcium looping tail-end process, although the process configuration is slightly different from what is presented in this paper, with the flue-gas to be treated extracted between two preheating stages and not downstream of the preheater. The difference in CAPEX, in $€/t_{clk}$, ranges from -5% to 11%, depending on the amount of CaO-rich purge from the calcium looping process that is added to the raw meal in the cement plant.

Considering membrane-based technologies, literature cost estimates for cement applications are difficult to compare with the results presented here, as they are based on different process concepts than considered in this work. For the CAP and integrated calcium looping technologies, detailed cost analyses comparable with the work presented here are not available in the literature.

Through the conduction of the current techno-economic analysis, several possibilities for improved cost performance or to reduce uncertainties in cost estimates have been identified.

For several technologies, it was observed that process contingencies contributed heavily to the CAPEX. The process contingencies account for costs that are unknown, and the relative amount of unknown costs are assumed to be higher for technologies with lower maturity (cf. Table 3). The process contingencies are particularly high for the oxyfuel, the MAL and the integrated CaL technologies, where they directly account for about 22%, 30% and 20% of the TDC for these technologies, respectively. In addition, elements of the fixed OPEX are calculated as a factor of the CAPEX, such that the process contingencies overall account for 14%, 20% and 15% of the CO_2 avoidance cost for the oxyfuel, MAL and the integrated CaL technologies. Increasing technology maturity by further development and demonstration of the technologies for cement applications would reduce the uncertainty of the costs, and possibly lower the overall cost estimates of the technologies.

The technical evaluation reported on in Part 1 showed that steam generation in a NG boiler contributed to a significant share of equivalent specific CO_2 emissions for the solvent technologies MEA and CAP. To reduce these emissions and potentially decrease the cost of CO_2 avoided, it could be considered to mix the flue gas from the NG boiler with those from the cement plant and thereby capture the CO_2 from the NG boiler in the post-combustion process. Furthermore, in cement plants which use raw material with low moisture content, a larger amount of waste heat would be available for steam generation at lower cost and lower associated CO_2 emissions compared to the NG boiler steam generation. This is for instance the case for the previously mentioned Norcem cement plant in Brevik, where it has been found that use of the plant waste heat for solvent regeneration can cover the heat demand for capturing ~40% of the emitted CO_2 [13]. Thus, the solvent based technologies will have a better techno-economic performance when retrofitted to such plants.

The oxyfuel technology is shown to have the lowest cost of CO_2 avoided, and has both relatively low OPEX and CAPEX, even though maturity related process contingencies do contribute significantly to the CAPEX. It should however be noted that there are several important aspects regarding retrofitability of the technology with the cement plant which could affect the cost process performance, such as potential impacts on the reliability of cement production due to substantial modifications of the core production process, that have not been considered in calculation of the economic KPIs.

For the MAL technology, CAPEX is the single largest cost factor. In this context, the membrane performance is essential as it strongly influences the energy performance of the whole process and consequently the size and cost of several of the most capital-intensive process equipment. The MAL process design was restricted to the specific membrane type that was tested within CEMCAP. A screening of different membranes, preferably with testing in real conditions at a cement plant to increase technology maturity and reduce uncertainties, in addition to further optimization of the system could result in better technical performance and lower costs than observed here.

For the calcium looping technologies, CAPEX and the consumption of coal are the largest cost factors, although in the tail-end configuration, the large coal consumption is effectively counterbalanced by the consequent production of electricity and the associated negative CO_2 avoidance costs (cf. Figure 5). In the integrated configuration, the CAPEX together with the capital-related fixed OPEX account for nearly 80% of the cost of CO_2 avoided. Further development of this technology on a larger scale is therefore essential to increase maturity, reduce uncertainties and potentially bring about cost reductions.

4.2. Sensitivity Analysis

Various assumptions on cost parameters are essentially dependent on the geographic location of the cement plant and the time at which the cost analysis is performed. The effect of this variability on the economic KPIs was investigated by varying the following parameters in the suggested ranges:

- Coal price: $+/-$ 50% of the reference cost
- Steam supply: $+/-$ 50% of the reference cost
- Electricity price: $+/-$ 50% of the reference cost
- CAPEX of CO_2 capture technologies: $+35/-15\%$
- Carbon tax: $0–100\ \text{€}/t_{CO2}$

The sensitivity of the cost of CO_2 avoided to the coal price, steam cost, electricity price and a change in CAPEX are shown in Figure 6 as well as the sensitivity of the cost of clinker to a carbon tax. The cost of coal affects the CaL processes, due to the significant increase in fuel consumption associated with the CaL technology. The MEA, CAP and MAL technologies are unaffected by the cost of coal since these technologies do not require additional coal consumption.

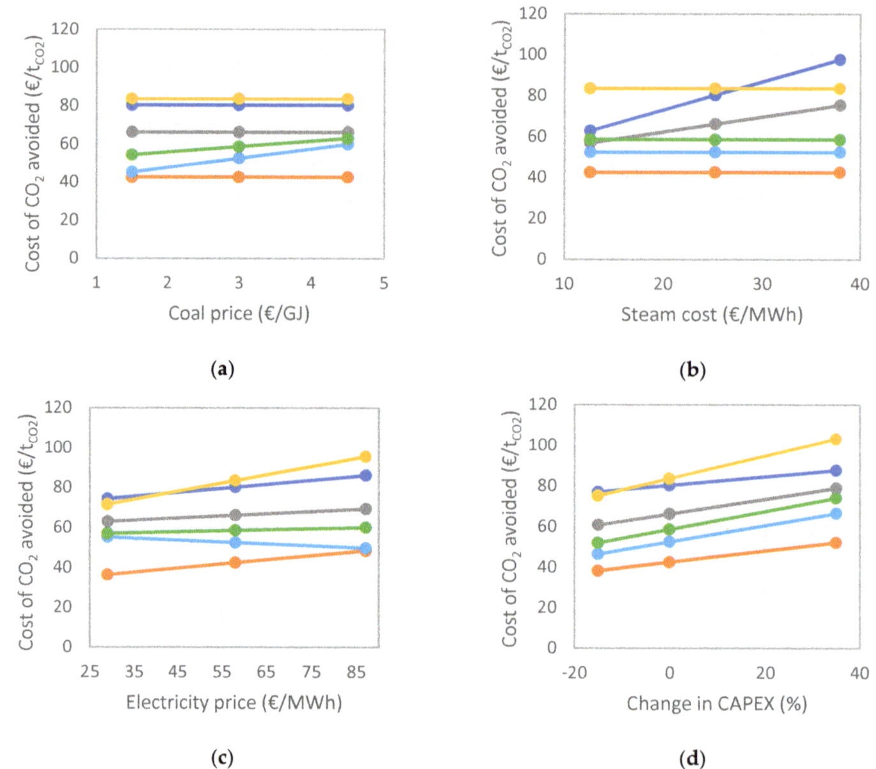

Figure 6. *Cont.*

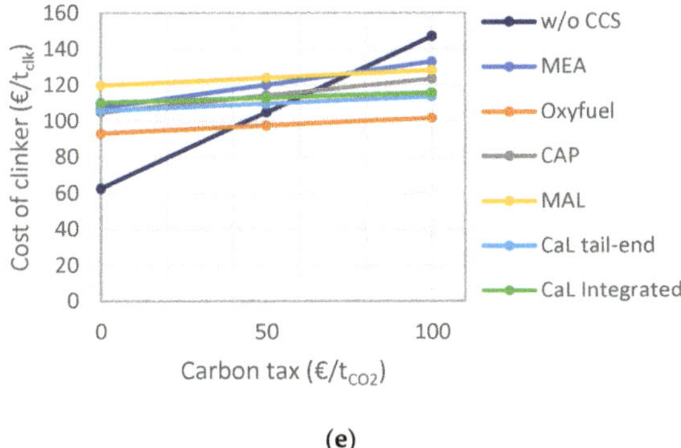

(e)

Figure 6. Sensitivity of the cost of CO_2 avoided to the (**a**) coal price, (**b**) cost of steam, (**c**) cost of electricity and (**d**) a change in CAPEX, and (**e**) sensitivity of the cost of clinker to a carbon tax.

The cost of steam naturally only affects the absorption-based MEA and CAP technologies, especially the MEA technology due to its relatively high steam requirement. At the lower end of the steam cost range, the cost of CO_2 avoided with MEA, CAP, and integrated CaL are almost the same.

Electricity intensive technologies, such as oxyfuel and MAL, are naturally the most sensitive to the price of electricity. The increase in electricity price decreases the cost of CO_2 avoided for the CaL tail-end technology, in contrast to all the other technologies. This is because the electricity generated in the CaL process covers a part of the cement plant's demand and therefore the CO_2 avoidance cost associated with electricity is negative for the CaL tail-end technology.

The most capital-intensive technologies, MAL and both CaL processes, are most sensitive to a change in the CAPEX estimate. The oxyfuel and CAP technologies are also significantly affected while the smallest effect is seen for MEA, which has the lowest CAPEX. It should be noted that the estimated fixed OPEX are also affected by a change in the CAPEX.

If a carbon tax is implemented on the direct CO_2 emissions from the cement plant, the cost of clinker for the reference cement kiln will increase. At a tax level of around 40 €/t$_{CO2}$, the cost of clinker (excluding costs for CO_2 transport and storage) with oxyfuel technology becomes lower than in the reference cement kiln, and at roughly 60 €/t$_{CO2}$ the CAP and both CaL technologies will have a lower cost of clinker compared with the cement kiln without CO_2 capture. For MEA and MAL, a carbon tax of around 75 €/t$_{CO2}$ would be required for a clinker cost lower than that of the reference cement kiln. Due to the direct CO_2 emissions from on-site steam generation for CO_2 capture with MEA and CAP, and therefore higher direct CO_2 emissions, these technologies are more sensitive to a carbon tax than the other CO_2 capture technologies.

4.3. Alternative Scenarios for CO_2 Capture

The results presented for the baseline scenario consider 90% CO_2 avoided from the cement plant flue gases, CO_2 transport by pipeline and, when required, steam being provided by a natural gas fired boiler. However, other scenarios for CO_2 capture have also been investigated within the CEMCAP project. This includes scenarios with higher CO_2 content in the flue gas, partial-scale capture, ship transport, different characteristics of the power generation system, steam import for solvent-based technologies, variations in air leakage in the oxyfuel cement plant and variations in the amount of sorbent purge used as raw material in the calcium looping tail-end configuration. Selected technology-specific scenarios which showed a significantly different composition or a change in the cost of CO_2 avoided are highlighted here, while the complete analysis of all scenarios can be found in the CEMCAP report by Voldsund et al. [39]. The cost of CO_2 avoided for the highlighted scenarios are presented in Figure 7 together with the cost of CO_2 avoided calculated for the baseline scenario for comparison (presented previously in Figure 5).

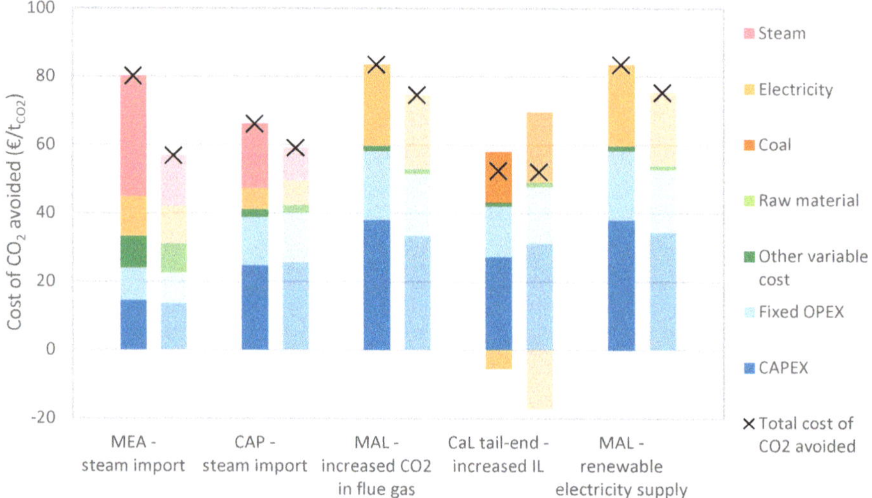

Figure 7. Cost of CO_2 avoided for alternative scenarios for CO_2 capture. The left-hand column shows the cost of CO_2 avoided calculated for the baseline scenario while the right-hand column shows the cost for the following technology-specific scenarios: MEA-steam import; CAP-steam import; MAL-increased CO_2 content in flue gases; CaL tail-end-reduced integration level.

For certain cement plants, it might be possible to import steam from an external coal-fired combined heat and power plant, instead of on-site generation from natural gas, to supply the MEA and CAP technologies. By doing so, the cost of steam could be reduced substantially [15] and consequently the cost of CO_2 avoided, as illustrated in the sensitivity analysis in Figure 6. Furthermore, depending on the power plant efficiency, the equivalent CO_2 avoided could be increased, leading to a further reduction of the specific cost of CO_2 avoided. The cost of CO_2 avoided for MEA and CAP when importing steam from a coal-fired CHP at a roughly 50% reduced cost and with around 20% lower CO_2 emissions per MWh$_{th}$ compared to steam from NG boiler [15] is shown in Figure 7. This results in 20% reduction in CO_2 avoided cost for MEA and 10% for CAP. The lower cost reduction for CAP compared with MEA is explained by CAPs significantly lower steam requirement. However, it should be mentioned that fewer than 10% of the existing cement plants in Europe are in close proximity to CHP plants.

An increased CO_2 content in the flue gas, which could be possible in a cement plant with e.g., increased maintenance to reduce air leak in the clinker burning line, was shown to benefit the CO_2 capture performance, and in particular the MAL technology. An increase in flue gas CO_2 content from the baseline scenario with an average of 20 mol% to a scenario with 22 mol% improves the process performance. In particular, the electricity requirement is reduced and the more efficient process results in reduced design capacity for most of the process equipment. As a result, the cost of CO_2 avoided is about 10% lower for the higher flue gas CO_2 content, 74.7 compared to 83.5 €/t$_{CO2}$ for the lower CO_2 content, as shown in Figure 7. Under these conditions, the MAL technology was found to outperform the reference technology MEA, which is not as strongly affected by the applied increase in CO_2 content of the flue gas (the cost of CO_2 avoided for MEA was found to decrease with <1%).

In the CaL tail-end configuration, the solid CaO-rich purge from the capture process is added to the raw meal in the cement kiln. The amount of Ca fed to the cement kiln from the sorbent purge to the total amount of Ca fed to the kiln is defined as the integration level (IL) between the tail-end calcium looping system and the kiln. The process presented in this paper has an IL of 50%. Designing for a lower IL will result in a larger potential for power generation from waste heat and could result in the cement plant being a net electricity producer with revenues for electricity export. The cost of CO_2 avoided for the Cal tail-end configuration when designed for 20% IL is shown in Figure 7. With this design, the CaL system requires significantly more fuel compared with the 50% IL design, but is also a net producer of electricity. This could be an important feature for a plant located in a region with high electricity prices and/or where the produced electricity substitutes generation with significantly higher specific CO_2 emissions. However, under the conditions applied in the cost analysis a similar balancing effect between the fuel consumption and electricity generation is seen in both designs and the cost of CO_2 avoided is calculated to be about the same, 52 €/t_{CO2}.

The characteristics of the power generation system in terms of efficiency and specific CO_2 emissions will depend on the geographical location of the cement plant and have an impact of the cost of CO_2 avoided, especially for electricity intensive technologies such as the MAL technology. The cost of CO_2 avoided for the MAL technology when electricity is generated solely from renewables, and with the same selling price is shown in Figure 7. The resulting CO_2 avoidance cost of 75.4 €/t_{CO2} is around 10% lower than calculated for the baseline scenario.

The investigation of alternative scenarios for CO_2 capture illustrates that the selection of a capture technology will depend strongly on plant-specific and local area characteristics, such as flue gas composition, vicinity to a potential steam exporter and electricity market conditions.

5. Conclusions

This paper presents a comparative cost assessment of CO_2 capture processes applied to a cement plant: MEA-based absorption as reference technology, chilled ammonia process, membrane-assisted CO_2 liquefaction, oxyfuel technology and calcium looping in a tail-end and an integrated configuration. Cost of clinker and cost of CO_2 avoided have been calculated based on detailed process simulations with input from experimental work and compilation of detailed equipment lists for each of the CO_2 capture technologies.

The cost analysis shows that the cost of clinker for the chilled ammonia and the calcium looping technologies is in the range of 105–110 €/t_{clk}, which is on the same level as the reference technology MEA. The oxyfuel technology has the lowest cost of clinker, 93 €/t_{clk}, and the membrane-assisted CO_2 liquefaction has the highest cost of clinker of 120 €/t_{clk}. Overall, the cost of clinker is shown to increase with 49–92% when CO_2 capture is retrofitted to the cement plant. The cost of CO_2 avoided lies between 42 €/t_{CO2} (oxyfuel process), which is approximately halved compared to MEA, and 84 €/t_{CO2} (membrane-assisted CO_2 liquefaction), which is on the same level as MEA.

The calculation of the economic KPIs relies on a number of assumptions related to important cost parameters which are dependent on location and time, such as cost of steam, electricity price and carbon tax. A sensitivity analysis showed the importance of such variables on the cost performance of the technologies. Further, the evaluation presented here is performed for application to a BAT reference cement kiln with steam generation primarily from natural gas. It should be noted that cement plants in general vary significantly from each other, for instance when it comes to CO_2 concentration in the flue gas, availability of waste heat or possibilities for importing steam from an external producer. The variability in these conditions was shown to have a strong impact on the economic performance of the CO_2 capture technologies, which indicates that the best CO_2 capture option in one cement plant might not be the best in another.

Part 1 of this paper series also showed that the characteristics of the power generation system, and the steam generation strategy, in terms of efficiency and specific CO_2 emissions, have a strong impact on the specific primary energy consumption and the equivalent CO_2 avoided. Furthermore, it was emphasized that several other aspects are important for evaluation and practical implementation of retrofitting technologies for CO_2 capture in a cement plant, such as technology maturity, integration with the clinker burning process and possible effects on product quality (and therefore risk), space requirement and the need for utilities, such as electric power or natural gas. It was found that the post-combustion technologies, MEA, chilled ammonia, membrane-assisted CO_2 liquefaction and calcium looping tail-end configuration are easier to retrofit than the more integrated technologies, oxyfuel and calcium looping integrated configuration.

The technologies investigated within CEMCAP are fundamentally different from each other and provide a portfolio of technologies with different properties, suitable for application in a wide variety of conditions in cement plants. No single technology has been found to stand out as a clear winner-each has its strengths and weaknesses. For the final selection of a CO_2 capture, a plant-specific techno-economic evaluation should be performed. In addition, plant-specific evaluation of more practical properties such as available space, capacity in local power grid and options for steam supply should also be carried out.

Supplementary Materials: The following are available online at http://www.mdpi.com/1996-1073/12/3/542/s1, Tables S1–S15: Equipment lists for CO_2 capture technologies.

Author Contributions: S.O.G. calculated the economic key performance indicators, and together with S.R. performed cost estimates for the MEA, MAL and oxyfuel technologies; E.D.L. and M.R. performed cost estimates, sizing and compilation of equipment lists for the CaL technologies; S.O.G., E.D.L., M.R., S.R. and M.V. contributed to defining the methodology, assembling cost data for non-standard process equipment and to the overall cost analysis; J.-F.P.-C., D.S., M.G. and M.M. contributed to sizing of equipment and compiling equipment lists for the CAP technology; D.B. contributed to sizing of equipment and compiling equipment lists for the MAL technology; C.F. contributed to sizing of equipment and compiling equipment lists for the MEA technology; R.A. contributed to sizing of equipment and compiling equipment lists for the oxyfuel technology; G.C. contributed to assembling cost data for non-standard process equipment and compiling equipment lists for the oxyfuel and CaL technologies; All authors contributed to reviewing and editing of the paper and this was coordinated by S.O.G. and M.V. S.O.G. wrote the paper.

Funding: This project has received funding from the European Union´s Horizon 2020 Research and Innovation Programme under Grant Agreement No. 641185, and the Swiss State Secretariat for Education, Research and Innovation (SERI) under contract number 15.0160.

Acknowledgments: The authors wish to thank Mr. Olaf Stallmann of Baker Hughes, a GE company for his contribution in direct cost estimation of the chilled ammonia process. Furthermore, the authors wish to thank Armin Jamali (VDZ), Helmut Hoppe (VDZ) and Kristin Jordal (SINTEF Energy Research) for their essential contribution to the technical analysis presented in Part 1 of this paper series.

Conflicts of Interest: The authors declare no conflict of interest.

Abbreviations

AACE	American Association of Cost Engineers
BAT	Best-Available Technologies
BHGE	Baker Hughes, a GE company
CAC	cost of CO_2 avoided
CaL	calcium looping
CAP	chilled ammonia process
CAPEX	capital costs
CHP	combined heat and power
COC	cost of clinker
CPU	CO_2 purification unit
CCS	carbon capture and storage
EC	equipment cost
ECRA	European Cement Research Academy
FEED	front-end engineering design
IC	installation cost
KPI	key performance indicator
MAL	membrane-assisted CO_2 liquefaction
MEA	monoethanolamine
OPEX	operating costs
ORC	organic rankine cycle

SNCR	selective non-catalytic reduction
TDC	total direct cost
TEC	total equipment cost
TPC	total plant cost

References

1. IEA; CSI. Technology Roadmap—Low-Carbon Transition in the Cement Industry. 2018. Available online: https://www.iea.org/publications/freepublications/publication/TechnologyRoadmapLowCarbonTransitionintheCementIndustry.pdf (accessed on 8 January 2019).
2. Garcia Moretz-Sohn Monteiro, J.; Goetheer, E.; Schols, E.; van Os, P.; Pérez-Calvo, J.F.; Hoppe, H.; Subrahmaniam Bharadwaj, H.; Roussanaly, S.; Khakharia, P.; Feenstra, M.; et al. Post-Capture CO_2 Management: Options for the Cement Industry (D5.1). 2018. Available online: https://www.sintef.no/globalassets/project/cemcap/presentasjoner/d5.1-final_rev1.pdf (accessed on 8 January 2019).
3. ECRA. CCS—Carbon Capture and Storage. Available online: https://ecra-online.org/research/ccs/ (accessed on 8 January 2019).
4. Bjerge, L.M.; Brevik, P. CO_2 capture in the cement industry, Norcem CO_2 capture project (Norway). *Energy Procedia* **2014**, *63*, 6455–6463. [CrossRef]
5. Knudsen, J.N.; Bade, O.M.; Askestad, I.; Gorset, O.; Mejdell, T. Pilot plant demonstration of CO_2 capture from cement plant with advanced amine technology. *Energy Procedia* **2014**, *63*, 6464–6475. [CrossRef]
6. Nelson, T.O.; Coleman, L.J.I.; Mobley, P.; Kataria, A.; Tanthana, J.; Lesemann, M.; Bjerge, L.M. Solid sorbent CO_2 capture technology evaluation and demonstration at Norcem's cement plant in Brevik, Norway. *Energy Procedia* **2014**, *63*, 6504–6516. [CrossRef]
7. Hägg, M.B.; Lindbråthen, A.; He, X.; Nodeland, S.G.; Cantero, T. Pilot Demonstration-reporting on CO_2 Capture from a Cement Plant Using Hollow Fiber Process. *Energy Procedia* **2017**, *114*, 6150–6165. [CrossRef]
8. Gassnova. CCS in Norway Entering a New Phase. 2018. Available online: https://www.gassnova.no/en/ccs-in-norway-entering-a-new-phase (accessed on 8 January 2019).
9. Liang, X.; Li, J. Assessing the value of retrofitting cement plants for carbon capture: A case study of a cement plant in Guangdong, China. *Energy Convers. Manag.* **2012**, *64*, 454–465. [CrossRef]
10. IEAGHG. Deployment of CCS in the Cement Industry; 2013/19. 2013. Available online: https://ieaghg.org/docs/General_Docs/Reports/2013-19.pdf (accessed on 8 January 2019).
11. Ozcan, D.C. Techno-Economic Study for the Calcium Looping Process for CO_2 Capture from Cement and Biomass Power Plants. Ph.D. Thesis, University of Edinburgh, Edinburgh, UK, 2014.
12. National Energy Technology Laboratory (NETL). Cost of Capturing CO_2 from Industrial Sources; DOE/NETL-2013/1602. 2014. Available online: https://www.netl.doe.gov/research/energy-analysis/search-publications/vuedetails?id=1836 (accessed on 20 November 2018).
13. Jakobsen, J.; Roussanaly, S.; Anantharaman, R. A techno-economic case study of CO_2 capture, transport and storage chain from a cement plant in Norway. *J. Clean. Prod.* **2017**, *144*, 523–539. [CrossRef]
14. Gerbelová, H.; van der Spek, M.; Schakel, W. Feasibility Assessment of CO_2 Capture Retrofitted to an Existing Cement Plant: Post-combustion vs. Oxy-fuel Combustion Technology. *Energy Procedia* **2017**, *114*, 6141–6149. [CrossRef]
15. Roussanaly, S.; Fu, C.; Voldsund, M.; Anantharaman, R.; Spinelli, M.; Romano, M. Techno-economic Analysis of MEA CO_2 Capture from a Cement Kiln—Impact of Steam Supply Scenario. *Energy Procedia* **2017**, *114*, 6229–6239. [CrossRef]
16. Rodríguez, N.; Murillo, R.; Abanades, J.C. CO_2 Capture from Cement Plants Using Oxyfired Precalcination and/or Calcium Looping. *Environ. Sci. Technol.* **2012**, *46*, 2460–2466. [CrossRef]
17. Diego, M.E.; Arias, B.; Abanades, J.C. Analysis of a double calcium loop process configuration for CO_2 capture in cement plants. *J. Clean. Prod.* **2016**, *117*, 110–121. [CrossRef]
18. Lindqvist, K.; Roussanaly, S.; Anantharaman, R. Multi-stage Membrane Processes for CO_2 Capture from Cement Industry. *Energy Procedia* **2014**, *63*, 6476–6483. [CrossRef]

19. Voldsund, M.; Anantharaman, R.; Berstad, D.; Cinti, G.; De Lena, E.; Gatti, M.; Gazzani, M.; Hoppe, H.; Martínez, I.; Monteiro, J.G.M.-S.; et al. *2018 CEMCAP Framework for Comparative Techno-Economic Analysis of CO$_2$ Capture from Cement Plants (D3.2)*. Available online: https://www.zenodo.org/record/1257112#.W8hidapPpaR (accessed on 8 January 2019). [CrossRef]

20. Jordal, K.; Voldsund, M.; Størset, S.; Fleiger, K.; Ruppert, J.; Spörl, R.; Hornberger, M.; Cinti, G. CEMCAP—Making CO$_2$ Capture Retrofittable to Cement Plants. *Energy Procedia* **2017**, *114*, 6175–6180. [CrossRef]

21. Anantharaman, R.; Berstad, D. Membrane and Membrane Assisted Liquefaction Processes for CO$_2$ Capture from Cement Plants. In Proceedings of the 14th International Conference on Greenhouse Gas Control Technologies, Melbourne, Australia, 21–25 October 2018.

22. Carrasco, F.; Grathwohl, S.; Maier, J.; Ruppert, J.; Scheffknecht, G. Experimental investigations of oxyfuel burner for cement production application. *Fuel* **2019**, *236*, 608–614. [CrossRef]

23. Jamali, A.; Fleiger, K.; Ruppert, J.; Hoenig, V.; Anantharaman, R. Optimised Operation of an Oxyfuel Cement Plant (D6.1). 2018. Available online: https://www.sintef.no/globalassets/project/cemcap/presentasjoner/d6.1-final_rev_1.pdf (accessed on 8 January 2019).

24. Trædal, S.; Berstad, D. Experimental Investigation of CO$_2$ Liquefaction for CO$_2$ Capture from Cement Plants (D11.2). 2018. Available online: https://www.sintef.no/globalassets/project/cemcap/2018-11-14-deliverables/d11.2-experimental-co2-liquefaction.pdf (accessed on 8 January 2019).

25. Pérez-Calvo, J.F.; Sutter, D.; Gazzani, M.; Mazzotti, M. Pilot tests and rate-based modelling of CO$_2$ capture in cement plants using an aqueous ammonia solution. *Chem. Eng. Trans.* **2018**, *69*, 145–150.

26. Pérez-Calvo, J.F.; Sutter, D.; Gazzani, M.; Mazzotti, M. Chilled Ammonia Process (CAP) Optimization and Comparison with Pilot Plant Tests (D10.3). 2018. Available online: https://www.sintef.no/globalassets/project/cemcap/2018-11-14-deliverables/d10.3_cap-optimization.pdf (accessed on 8 January 2019).

27. Alonso, M.; Álvarez Criado, Y.; Fernández, J.R.; Abanades, C. CO$_2$ Carrying Capacities of Cement Raw Meals in Calcium Looping Systems. *Energy Fuels* **2017**, *31*, 13955–13962. [CrossRef]

28. Arias, B.; Alonso, M.; Abanades, C. CO$_2$ Capture by Calcium Looping at Relevant Conditions for Cement Plants: Experimental Testing in a 30 kWth Pilot Plant. *Ind. Eng. Chem. Res.* **2017**, *56*, 2634–2640. [CrossRef]

29. Turrado, S.; Arias, B.; Fernández, J.R.; Abanades, J.C. Carbonation of Fine CaO Particles in a Drop Tube Reactor. *Ind. Eng. Chem. Res.* **2018**, *57*, 13372–13380. [CrossRef]

30. Hornberger, M.; Spörl, R.; Scheffknecht, G. Calcium Looping for CO$_2$ Capture in Cement Plants—Pilot Scale Test. *Energy Procedia* **2017**, *114*, 6171–6174. [CrossRef]

31. De Lena, E.; Spinelli, M.; Romano, M.; Gardarsdottir, S.O.; Roussanaly, S.; Voldsund, M. CEMCAP Economic Model Spreadsheet. 2018. Available online: https://zenodo.org/record/1446522 (accessed on 8 January 2019).

32. Anantharaman, R.; Bolland, O.; Booth, N.; van Dorst, E.; Ekstrom, C.; Fernandes, E.S.; Franco, F.; Macchi, E.; Manzolini, G.; Nikolic, D.; et al. European Best Practise Guidelines for Assesment of CO$_2$ Capture Technologies. D1.4.3 in DECARBit Project. 2011. Available online: https://www.sintef.no/globalassets/project/decarbit/d-1-4-3_euro_bp_guid_for_ass_co2_cap_tech_280211.pdf (accessed on 8 January 2019).

33. Cinti, G.; Anantharaman, R.; De Lena, E.; Fu, C.; Gardarsdottir, S.O.; Hoppe, H.; Jamali, A.; Romano, M.; Roussanaly, S.; Spinelli, M.; et al. Cost of Critical Components in CO$_2$ Capture Processes (D4.4). 2018. Available online: https://www.sintef.no/globalassets/project/cemcap/2018-11-14-deliverables/d4.4-cost-of-critical-components-in-co2-capture-processes.pdf (accessed on 8 January 2019).

34. NETL. Quality Guidelines for Energy System Studies: Cost Estimation Methodology for NETL Assessments of Power Plant Performance. 2011. Available online: https://www.netl.doe.gov/File%20Library/research/energy%20analysis/publications/QGESSNETLCostEstMethod.pdf (accessed on 20 November 2018).

35. Deng, H.; Roussanaly, S.; Skaugen, G. Techno-economic analyses of CO$_2$ liquefaction: Impact of product pressure and impurities. *Int. J. Refrig.* **2018**. submitted for publication.

36. Roussanaly, S. Calculating CO$_2$ avoidance costs of carbon capture and storage from industry. *Carbon Manag.* **2018**. [CrossRef]

37. De Lena, E.; Spinelli, M.; Gatti, M.; Scaccabarozzi, S.; Consonni, S.; Cinti, G.; Romano, M. Techno-economic analysis of Calclium Looping processes for low CO$_2$ emission cement plants. *Int. J. Greenh. Gas Control* **2019**, *82*, 244–260. [CrossRef]

38. IEAGHG. Cost of CO_2 Capture in the Industrial Sector: Cement and Iron and Steel Industries. 2018-TR03. 2018. Available online: http://documents.ieaghg.org/index.php/s/YKm6B7zikUpPgGA?path=%2F2018%2FTechnical%20Reviews (accessed on 8 January 2019).
39. Voldsund, M.; Anantharaman, R.; Berstad, D.; De Lena, E.; Fu, C.; Gardarsdottir, S.O.; Jamali, A.; Pérez-Calvo, J.F.; Romano, M.; Roussanaly, S.; et al. CEMCAP Comparative Techno-Economic Analysis of CO_2 Capture in Cement Plants (D4.6). 2018. Available online: https://www.sintef.no/globalassets/project/cemcap/2018-11-14-deliverables/d4.6-cemcap-comparative-techno-economic-analysis-of-co2-capture-in-cement-plants.pdf (accessed on 8 January 2019).

 *energies*

Article

A Study on the Evolution of Carbon Capture and Storage Technology Based on Knowledge Mapping

Hong-Hua Qiu [1,2,*] and Lu-Ge Liu [1]

[1]   School of Law, Northwest University, Xi'an 710127, China; liuluge@stumail.nwu.edu.cn
[2]   Max Planck Institute for Innovation and Competition, Marstallplaz 1, 80539 Munich, Germany
*   Correspondence: qiuhonghua@nwu.edu.cn or Honghua.Qiu@ip.mpg.de

Received: 3 April 2018; Accepted: 20 April 2018; Published: 1 May 2018

Abstract: As a useful technical measure to deal with the problem of carbon dioxide (CO_2) emissions, carbon capture and storage (CCS) technology has been highly regarded in both theory and practice under the promotion of the Intergovernmental Panel on Climate Change (IPCC). Knowledge mapping is helpful for understanding the evolution in terms of research topics and emerging trends in a specific domain. In this work knowledge mapping of CCS technology was investigated using CiteSpace. Several aspects of the outputs of publications in the CCS research area were analyzed, such as annual trends, countries, and institutions. The research topics in this particular technology area were analyzed based on their co-occurring keyword networks and co-citation literature networks, while, the emerging trends and research frontiers were studied through the analysis of burst keywords and citation bursts. The results indicated that the annual number of publications in the research field of CCS technology increased rapidly after 2005. There are more CCS studies published in countries from Asia, North America, and Europe, especially in the United States and China. The Chinese Academy of Sciences not only has the largest number of publications, but also has a greater impact on the research area of CCS technology, however, there are more productive institutions located in developed countries. In the research area of CCS technology, the main research topics include carbon emissions and environmental protection, research and development activities, and social practical issues, meanwhile, the main emerging trends include emerging techniques and processes, emerging materials, evaluation of technological performance, and socioeconomic analysis.

Keywords: carbon capture and storage; knowledge mapping; technological evolution; CiteSpace

---

1. Introduction

1.1. The Significance of Carbon Capture and Storage Technology

As one of the most significant components of long-lived anthropogenic greenhouse gases (GHGs), carbon dioxide (CO_2) may contribute to long-term global warming. During the period of 1959–2008, on average, the CO_2 emissions were about 43% of the total emissions every year. In order to control global warming, it is essential to achieve the stabilization of CO_2 concentrations [1]. To capture and sequestrate the carbon dioxide which is called carbon capture and storage (CCS) technology, has been regarded as an effective way to reduce the emissions of CO_2 [2]. CCS technology is really a set of technologies, which normally consists of four elements, which are capture, compression, transportation, and storage. This significant technology can not only decrease emissions of carbon dioxide generated from the utilization of fossil-based energy by more than 65% [3], but also provide an increasingly-important hedge for fossil fuels energy; moreover, this technology can responsible for about 16.67% of total reductions of carbon dioxide emissions required by 2050 [4]. As a promising technology to deal with global warming, CCS technology will be of great importance to decrease the CO_2 emissions [5].

Since at least 2005, when a special investigation on CCS technology was released by the IPCC [6], international institutions (International Energy Agency, IEA) and some developed countries in North America (i.e., the United States and Canada) and Europe (i.e., the United Kingdom and Norway) have begun to invest in and deploy CCS projects, and these projects include eight large-scale integrated projects around the world in 2013 and 14 large-scale projects in 2015 [7].

1.2. The Importance and Application of Knowledge Mapping

Against the backdrop of the knowledge economy, knowledge has become the most valuable parts of intangible assets for most entities [8], since the effective management of knowledge will be favorable to the development of science and technology [9]. Organizations constantly utilize their own knowledge to seek chances to improve their innovative activities. Knowledge management can be an effective measure to deploy knowledge assets [10]. The capabilities of knowledge management could help a firm improve its efficiency in using resources and the performance of technological innovation [11]. Knowledge management can contribute to understanding the situation and gaps of different researchers/research institutions, and define agendas for future research [12].

However, there is usually a knowledge gap between the knowledge what has been already known and the knowledge in practical application, and this gap is particularly important and should be faced when a company wants to develop a new product or process [13]. It is necessary for companies to know the mechanisms whereby knowledge is created and how knowledge can be used at different levels, such as technological R&D, institution and policy, and more. Therefore, knowledge mapping could help researchers or companies to know the situation and trends of knowledge flow and identify the existing gaps [14]. Knowledge mapping, also called knowledge representation, is a process in which a schematic representation of knowledge is created [15]. During the activities of knowledge management in an organization, knowledge mapping is a starting point of knowledge management. It can be a significant tool to obtain and communicate the explicit knowledge, meanwhile, it also can serve as a pointer to the organizations who own the implicit knowledge [16]. In recent years, knowledge mapping has shown increasing interest in the topic of knowledge management [8].

Scientometrics is one of the ways to realize knowledge mapping. It is a quantitative analysis about a process, as well as input and output of scientific and technological activities to understand the knowledge mapping of scientific fields and developing statuses in relevant scientific and technological domains based on the methods of computing technology and social statistics [17]. In recent years, knowledge mapping has been investigated using scientometrics. Mercuri et al. [18] applied the scientometric approach to show the global distribution of microbial fuel cell technology and uncover the scale of application of the technology. Montoya et al. [19] used bibliometric analysis techniques to investigate the current situation and future trends, as well as dynamic changes in the energy field in Spain. Konur [20] assessed the main developing trends and problems in the technological field of the algae and bio-energy based on an investigation of scientometric analysis.

According to the existing literature the number of publications in the CCS research area has experienced a continuous and stable growth. Meanwhile, given the great importance of knowledge mapping, it is urgent and necessary to acquire a correct understanding of the current situation and developing trends, as well as the hot spots and fronts of CCS research area by using the analysis tool of knowledge mapping, which is the objective of this paper. This paper will be organized as follows: Section 2 describes data collection and the knowledge mapping research software, Section 3 analyzes the results. The discussion is described in Section 4. Section 5 summaries the conclusions, and finally, Section 6 presents the research limitations.

2. Methodology

2.1. Knowledge Mapping and CiteSpace

With the rapid popularity and development of Information Technology (IT), is has become of great significance in knowledge management activities [21]. In this paper, CiteSpace was employed as a useful tool to obtain the information about knowledge mapping of CCS technology. CiteSpace is a freely available Java application software, developed by the Chen research group at Drexel University (V. 5. 0. R7, SE, 32-bit, Philadelphia, PA, USA). It was designed as an effective tool for mapping the knowledge of scientific literature and helping researchers to find the critical points in a specific area. Furthermore, some special analysis functions related to concepts of centrality of CiteSpace, such as detection analysis of bursts and in-between centrality analysis [22], will be used to visualize knowledge mapping of CCS technology, including analyzing the distribution of documents, identifying the research frontiers, detecting the emerging trends and abrupt changes, and more.

Since CiteSpace became available on 13 September 2004, not only the software service provider [23], but also many other researchers have used CiteSpace to carry out studies regarding knowledge management and knowledge mapping. Yu and Xu [24] used CiteSpace to study the status quo and developing trends in the research area of carbon emission trading. Chen and Guan [25] applied CiteSpace to investigate general development and research performance of emerging nano-biopharmaceuticals. Based on CiteSpace, Lin, Wu, and Hong [26] studied hotspots and future trends of ecological assets research.

2.2. Data Collection

To investigate the evolution of CCS technology from the perspective of knowledge mapping of CCS studies is the major objective of this paper. Literature collection is the essential basis for knowledge mapping. The bibliometric data were searched from the Web of Science in this study. The bibliometric search strategy can be described as follows. The timespan was "all years", the database was "Web of Science Core Collection", the topic was "carbon capture and storage technology". As seen in Figure 1, there are two groups of keywords in the search terms. "Group 1" is the keywords related to "CO_2", including "carbon", "CO_2" and "carbon dioxide". "Group 2" is the keywords related to "capture and storage", such as "captur*", "storag*", "sorbent", "transportation", "adsorption", "separation", and "adsorbent". The literature will be considered as our research sample if its title includes any keyword in both "Group 1" and "Group 2". All the studies were converted into a specific format to meet the requirements of CiteSpace software. It should be noted that, taking the accuracy and relativity of the sampled literature into account, this paper only collected literature using the "title", but not the "abstract" or "keywords". However, it should be admitted that the research studies, to some extent also reflected the research situation of CCS technology and had the academic value of their own in the research area of CCS technology if their "abstract" or "keywords" appeared the retrieval terms of this research topic. However, these articles had little impact on the study results.

Under the retrieval conditions and strategy described above, a total number of 10,180 CCS publications were searched on 26 January 2018. There are 12 various literature types in the final database, such as articles (9686), meeting abstracts (610), proceeding papers (533), reviews (385), news items (190), editorial materials (167), and corrections (56), as well as letters (44), and more. Among them, the major literature type is article, which accounted for 94.74% of the total literature. Taking into account the validity and representation of the literature, and in order to eliminate the "information noise" of the database, articles were included in further investigation and the other types of literature were excluded. The flow diagram of sample literature selection process is presented in Figure 1.

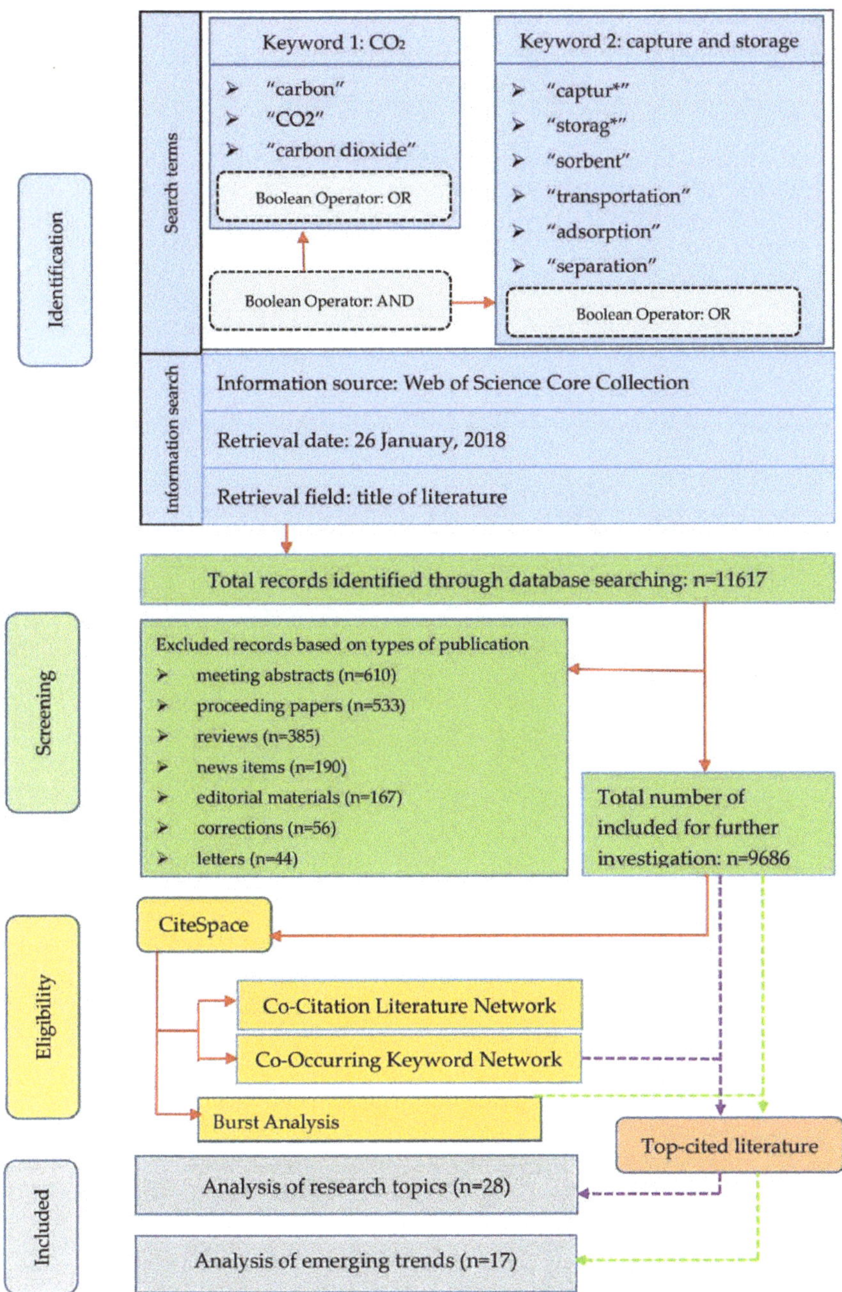

Figure 1. Process of literature retrieval and selection.
Figure 1. Process of literature retrieval and selection.

3. Analyses and Results

3.1. Basic Analysis of Publication Outputs

3.1.1. Annual Trend Analysis

The annual total publications during the study period is shown in Figure 2. The annual number of articles in the study area of CCS technology increased steadily from 60 in 2000 to a peak of 1372 in 2016. Additionally, a polynomial regression equation was employed to describe the relationship between literature publications and the year of publishing. The regression equation is $y = 1.6511x^2 + 25.187x$, where x represents the number of year since 2000 and y represents the publication number. During the given time period, the number evolution of published articles can be split into two parts. The first period is before 2006. The annual number of documents in the research area of CCS technology increased gradually and fluctuated during this time. Correspondingly, the second period is after 2006, and the annual number of publication in the research field of CCS technology increased rapidly. It should be noted that, because the terminal date for downloading literature was 26 January 2018, and there is a time lag for literature to be indexed by a database, the data for 2017 and, especially, 2018 would be incomplete.

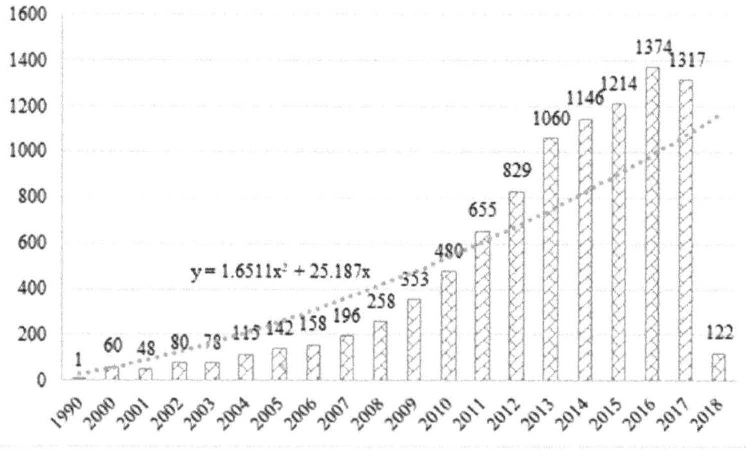

Figure 2. Annual number of publications in the carbon capture and storage (CCS) technology area.

3.1.2. Country Analysis

The different number of publications in every country reflects to some extent the different degrees of concern regarding the research on CCS technology. Table 1 lists the top 20 most productive countries in the research area of CCS technology. It can be seen that China ranked first, with 2455 articles, accounting for 25.356% of the total publications, which was followed by the USA with 2197 articles accounting for 22.682% of the total. Other productive countries included the UK (655), South Korea (600), and Australia (578), as well as Canada (509). Moreover, among the top 20 countries, there are nine European countries, seven Asian countries, two North American countries, one South American country and one Oceania country.

Table 1. Top 20 most productive countries in carbon capture and storage (CCS) research area.

Countries	Number of Publications	Percentage of Total Number	Countries	Number of Publications	Percentage of Total Number
China	2455	25.356%	Japan	322	3.324%
USA	2197	22.682%	Italy	306	3.159%
UK	853	8.807%	Norway	301	3.108%
South Korea	600	6.195%	Netherlands	292	3.015%
Australia	578	5.967%	Sweden	201	2.075%
Canada	509	5.255%	Iran	198	2.044%
Germany	476	4.914%	Poland	152	1.569%
Spain	437	4.512%	Malaysia	152	1.569%
France	381	3.934%	Singapore	146	1.507%
India	328	3.386%	Brazil	141	1.456%

Note: the number of publications of China contains 155 articles from Taiwan and the number of publications of the UK contains 198 articles from Scotland.

3.1.3. Institution Analysis

The main researchers and their research capabilities in a specific domain can be identified from the statistical and analytical perspectives of publications. Table 2 lists the top 20 most productive institutions in the research field of CCS technology. What can be seen is that the Chinese Academy of Sciences of China (CAS of China) had the most CCS articles, with 514 publications, followed by another Chinese institution, Tsinghua University, with 122 publications, and the third and fourth places are the institutions from the United States, the University of California, Berkeley (UC Berkeley, 118) and the U.S. Department of Energy (DOE, 105). Among the most productive institutions listed in Table 2, there are six institutions from China, five institutions from the U.S., and three institutions from the United Kingdom.

Table 2. Top 20 most productive institutions in the CCS research area.

Institution	Country	Number of Publications	Freq.	Centrality
Chinese Academy of Sciences	China	514	431	0.35
Tsinghua University	China	122	121	0.06
University of California, Berkeley	USA	118	109	0.15
U.S. Department of Energy	USA	105	100	0.05
Tianjin University	China	99	92	0
Georgia Institute of Technology	USA	98	92	0.02
University of Texas at Austin	USA	98	94	0.05
Zhejiang University	China	98	92	0.02
Consejo Superior de Investigaciones Cientificas (CSIC)	Spain	96	84	0.01
Norwegian University of Science and Technology	Norway	95	87	0.03
National University of Singapore	Singapore	93	88	0.03
Monash University	Australia	87	86	0.04
University of Queensland	Australia	87	79	0.01
The University of Edinburgh	UK	86	78	0.08
Imperial College of Science, Tech & Medicine	UK	86	80	0.08
University of Nottingham	UK	82	76	0.02
Oak Ridge National Laboratory	USA	80	63	0.08
The University of Melbourne	Australia	78	67	0.03
Huazhong University of Science and Technology	China	77	75	0.01
Southeast University	China	76	68	0

Note: the number of publications of Chinese Academy of Sciences contains 78 articles from the University of Chinese Academy of Sciences.

In terms of frequency, the CAS of China is at the top with 431, and then comes Tsinghua University (121), UC Berkeley ranks the third with 109. These institutions with high cited frequency had a greater impact on the research field of CCS technology. In terms of centrality, the top five institutions were the Chinese Academy of Sciences (0.35), the University of California, Berkeley (0.15), and the University of Edinburgh (0.08), the Imperial College of Science, Tech & Medicine (0.08), and Oak Ridge National Laboratory (0.08). Institutions with high centrality means that these institutions are significant in

Energies **2018**, *11*, 1103

and the University of Edinburgh (0.08), the Imperial College of Science, Tech & Medicine (0.08), and Oak Ridge National Laboratory (0.08). Institutions with high centrality means that these institutions are significant in CCS research area. It is worth noting that the institutions from developed countries have a higher value in both cited frequency and centrality, which implies that the institutions of developed countries discovered more significant technologies and have stronger innovative capabilities in this specific domain. Generally speaking, the institutions listed in the table paid more attention to the research on CCS technology, obtained more scientific works, and had a relative better research foundation. Figure 3 shows the co-occurring network of institutions in the domain of CCS technology. The different size of circle nodes represent different number of documents published by each institution and the larger the note size of the institution, the more literature it published, for example, the circle nodes of CAS of China, DOE of the U.S., Monash University, and CSIC, etc., are obvious larger in the figure and these institutes also published more articles in the domain of CCS technology. Moreover, it also can be seen that among the most significant institutions, they were linked tightly each other, which led to more important notes (institutions) having a higher centrality. This reflects not only the structure and position of research institutions in the domain of CCS technology, but also the collaboration pattern for productive institutions.

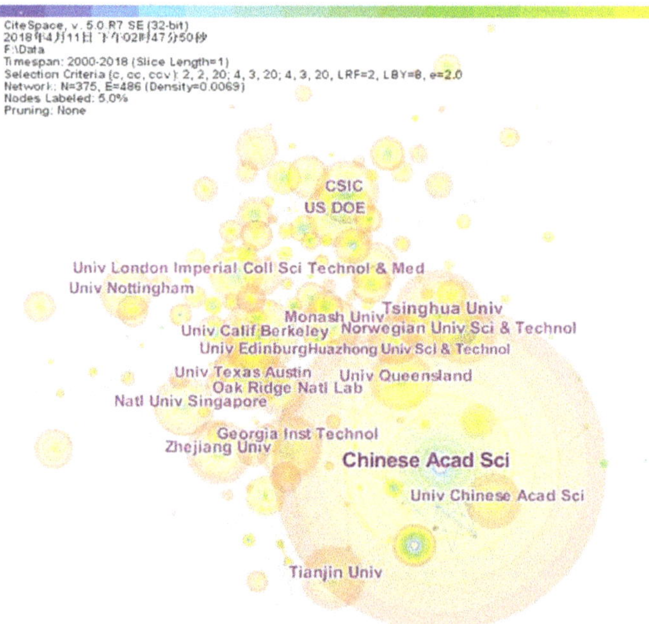

Figure 3. Mapping on the co-occurring network of institutions.

3.2. Research Topics Analysis

3.2.1. Co-Citation Literature Network

Literature analysis is one of the primary contents of knowledge mapping. The literature references can be analyzed from the perspective of visualization. The network of the literature co-citation in the research area of CCS technology can be developed using CiteSpace, which is shown in Figure 4. The nodes are the literature topics and the lines illustrated as various colours in the figure represent that the co-citation frequency of lines reached the threshold. What is shown that the important publications were tightly linked, which means that those studies concerned about similar research topics. Some previous studies represent the major research contents in this specific domain, while the core research concepts received a progressive and continuous development. Furthermore,

core research concepts received a progressive and continuous development. Furthermore, the nodes separated from the core literature represent some new research branches, especially the new technical means and practice of this particular area.

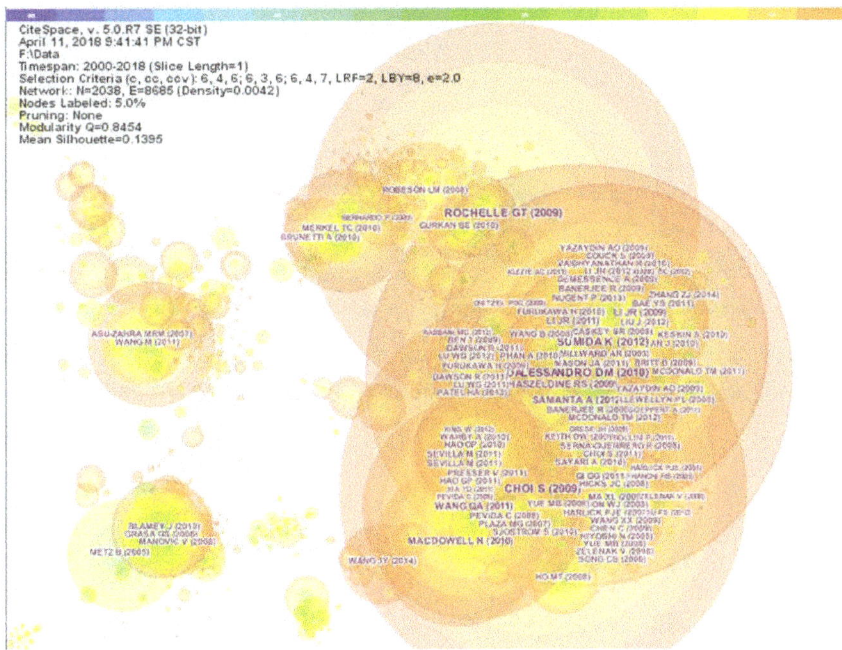

Figure 4. Co-citation literature network of CCS research area.

Table 3 lists the Top 20 references in CCS research area according to their frequency of citation. They were selected for analysis by CiteSpace. It can be seen that the literature was the most frequently-cited article and the frequency was 656, in which latest advances and developing trends in CO_2 separation were discussed, and a particular attention was also paid on advances in the technical area of metal-organic frameworks [27]. Then, the literature ranked 2nd has a frequency of 625, in which CO_2 adsorption behaviors were described while the current available CO_2 adsorbents and their significant features were also presented [28].

In terms of the source of the highly cited literature in the research area of CCS technology, there were three documents published in *Energy & Environmental Science* (Impact factor: 29.518). In addition, there are two documents published in *Science*. One of the references was published in *Science*, with a frequency value of 663, in which the process and solvent improvements of amine scrubbing to separate CO_2 from coal-fired power plants were investigated [29]. The other article published in *Nature* with frequency value of 180, in which the technological applications about the metal-organic materials (MOMs) were reported [30]. The only one book listed in the table is edited by Metz B. [31] and published in 2005. This book has important influences on both academic and practice in CCS domain. It not only described the situation and challenges of carbon dioxide emissions, analyzed the technological contents of CCS, but also discussed the problems related to the technical applications of CCS.

Table 3. Top 20 references in the research area of CCS technology.

Energies **2018**, *11*, 1103

Authors	Literature Information	Freq.	Year
D'Alessandro DM	Angew Chem Int Edit, V49, P6058	656	2010
Choi S	ChemSusChem, V2, P796	625	2009
Rochelle GT	Science, V325, P1652	603	2009
Sumida K	Chem Rev, V112, P724	505	2012
Wang QA	Energ Environ Sci, V4, P42	347	2011
Figueroa JD	Int J Greenh Gas Con, V2, P9	329	2008
Samanta A	Ind Eng Chem Res, V51, P1438	299	2012
Yang HQ	J Environ Sci-China, V20, P14	278	2008
Haszeldine RS	Science, V325, P1647	272	2009
Macdowell N	Energ Environ Sci, V3, P1645	245	2010
Li JR	Coordin Chem Rev, V255, P1791	225	2011
Li JR	Chem Soc Rev, V38, P1477	209	2009
Millward AR	J Am Chem Soc, V127, P17998	195	2005
Merkel TC	J Membrane Sci, V359, P126	192	2010
Wang M	Chem Eng Res Des, V89, P1609	191	2011
Caskey SR	J Am Chem Soc, V130, P10870	190	2008
Hao GP	Adv Mater, V22, P853	190	2010
Boot-Handford ME	Energ Environ Sci, V7, P130	187	2014
Nugent P	Nature, V495, P80	180	2013
Metz B	Book	180	2005

Note: Taking the large number of literature into account, we just selected the Top 20 references for analysis.

3.2.2. Co-Occurring Keyword Network

The keywords can reflect the important concepts or core contents of a document. The research topics and hotspots can be detected using the keyword co-word analysis of CiteSpace [32]. The position, collaboration, and connection of the keywords of the references in the research area of CCS technology can be investigated using a keyword co-occurring network map. Figure 5 shows the co-occurring network using keywords as "node type".

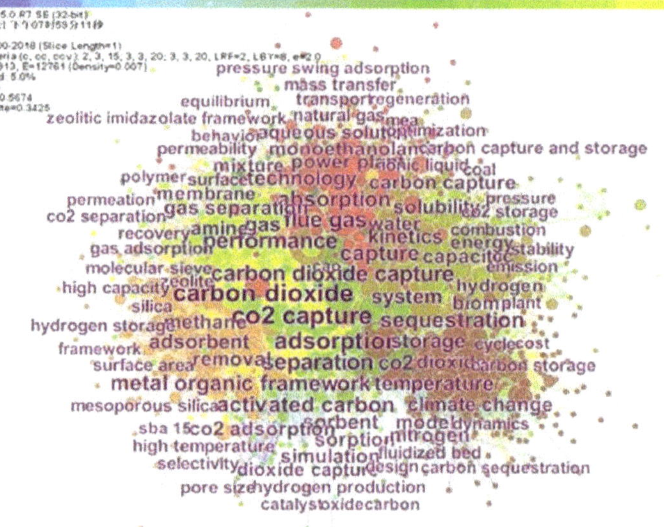

Figure 5. Co-occurring keyword network of CCS research area.

It can be seen that most of the keywords are closely connected to each other. Therefore, it can be concluded that the main studies in the field of CCS technology were interlinked, the core keywords in this specific domain includes "carbon dioxide", "CO_2 capture", "adsorption", "separation", "performance", "metal organic framework", and "activated carbon" as well as "adsorption", and more. It means that these keywords represent the research topics in the field of CCS technology. In addition, some keywords, such as "hydrogen storage", "membrane", "polymer", "permeability", "pressure", and "carbon sequestration", have become the important research branches in the research area of CCS technology. Although these keywords have a weaker correlation with core keywords, they may also, to some extent, represent novel theories or practical requirements in the field of CCS technology.

In order to investigate the popular research topics in the discipline of CCS technology and their evolution over time, the keywords that appeared more than 300 times were collected and listed based on years in Table 4. A total of 35 keywords are listed in the table. Because the purpose of this analysis is to investigate the changes in centrality over time, words/phrases with the same meaning were not merged in this study. As seen in the table, in terms of frequency, "carbon dioxide" appeared most frequently of all the keywords, followed by "CO_2 capture", "adsorption", "separation", and "carbon dioxide capture".

Table 4. Main research topics by year.

Year	Keyword	Centrality	Freq.
2000	Carbon dioxide	0.06	1921
2001	CO_2 capture	0.11	1866
2000	Adsorption	0.1	1375
2000	Separation	0.07	1055
2009	Carbon dioxide capture	0.01	1048
2006	Performance	0.03	847
2002	Flue gas	0.04	813
2009	Metal organic framework	0.03	800
2003	Capture	0.07	736
2002	Absorption	0.03	729
2000	Activated carbon	0.05	682
2000	Sequestration	0.11	670
2000	CO_2	0.17	621
2005	Adsorbent	0.02	582
2007	Technology	0.01	533
2000	Temperature	0.08	531
2000	System	0.08	514
2002	Storage	0.04	488
2001	Kinetics	0.12	485
2005	Sorbent	0.03	458
2002	Gas	0.04	452
2000	Dioxide	0.1	438
2000	Removal	0.03	421
2000	Water	0	390
2001	Sorption	0.02	388
2005	CO_2 adsorption	0.02	373
2010	Carbon capture	0	371
2003	Power plant	0.02	367
2004	Simulation	0.04	367
2002	Gas separation	0.04	348
2002	Capacity	0.03	315
2004	Amine	0.01	313
2006	Solubility	0.07	309
2000	Climate change	0.04	305
2007	Monoethanolamine	0	303

The studies represented by keywords with a highly-cited frequency have been concerned by many institutions/scientists. Therefore, these keywords can represent the research hotspots in the research area of CCS technology. In addition, it also indicates that the academic researches on CCS technology are mainly due to people's attention to the problems of CO_2 and climate change. Some specialized disciplines included in the table, such as "kinetics", "flue gas", "adsorbent", "power plant", and "metal organic framework", reflect the continuous development of CCS technology and the expansion of its application areas.

In terms of centrality, "CO_2" is the most important keyword with a centrality of 0.17, which is quite a high centrality value among all the keywords listed in the table. It is followed by "Kinetics" with a centrality of 0.12, and then "CO_2 capture" and "sequestration", both of which have a centrality of 0.11. This means that the researches represented by the keywords with high centrality have a greater impact on the research area of CCS technology, therefore, they represent the core contents and important sectors in this specific domain of technology.

The valuable information from the changes of keywords over time also can be analyzed. In order to illustrate the developing trends in CCS research area, the keywords listed in Table 4 can be divided into two periods, 2000–2005 and 2006–present. It can be seen that there are 28 keywords in the period of 2000–2005, which accounts for 80% of the total keywords and seven keywords in the period of 2006–present. The keywords "Carbon dioxide capture" in both of the two periods were cited frequently. It can be considered that the research contents represented by the keywords have not changed for a long period and became the foundation of the research field of CCS technology. Some keywords, such as "performance", "metal organic framework", "technology", "solubility", and "monoethanolamine", did not appear in the period between 2000–2005, but they appeared in the period from 2006–present, which shows that these keywords may represent new research topics related to practical applications in the domain of CCS technology in recent years.

3.2.3. Main Contents of Important Research Topics

(1) Representative significant research contents related to "Carbon dioxide" or "CO_2" (2000).

In the study by Hobbie et al. [33], they believed that although arctic and boreal area is of significance in current carbon circle, carbon sequestration activities within northern soils have not been understood very well, to explore the special properties of soil carbon in the systems of high latitude can improve current knowledge about carbon fluxes and even be helpful for evaluating the impacts of these systems on climate changes in the future. Schimel et al. [34] used the historical climate data to invest the impacts of CO_2 and climate on the carbon sequestration in the U.S. ecosystems. Treseder et al. [35] discussed that it is very important for evaluating the effect of world climate change on the circling of mycorrhizal carbon to investigate the decomposition of hyphal under the conditions of elevated carbon dioxide and nitrogen. Ravikovitch et al. [36] considered that carbonaceous adsorbents can play significant roles in practice application, therefore, they tried to investigate a special method to obtain the feature of pore size of micro-porous carbonaceous substances. Ding and Alpay [37] carried out an experiment and found that the complexities of carbon dioxide adsorption on hydrotalcite could change the adsorption efficiency of the material.

(2) Representative significant research contents related to "kinetics" (2001).

Sobkowski and Czerwiński [38] reported the kinetics of carbon dioxide adsorption by using the technical method of radiometry. They found that the adsorption feature of carbon dioxide is because the surface reaction on the electrode, the adsorption rate will raise if the electrode is set under a specific range, the concentration of surface will not be affected by the temperature and carbon dioxide concentration. Nugent et al. [30] examined how, in order to purify the products (i.e., fresh water and gases) in the industrial field, the expenditure of energy utilization could account for about 15% of energy output worldwide, and the request for the products is expected to increase three times by the year 2050. They described that the metal-organic materials with special technical properties offered an unprecedented carbon dioxide sorption selectivity over nitrogen, hydrogen and methane. Ding and Alpay [37] reported that, since the 1950s, the kinetics of carbon dioxide adsorption by different kinds of adsorbents had been examined broadly, while carbon dioxide adsorption capacity on hydrotalcite have been present recently. Serna-Guerrero and Sayari [39] reported that greenhouse gases have serious implications for our environment and ecosystem, lots of initiatives and new technologies were developed to deal with the emissions of carbon dioxide, and therefore, they tried to simulate the CO_2 adsorption kinetic model in their study. Ochoa-Fernandez et al. [40] described that, how to remove the carbon dioxide from waste gases is increasingly important during the activities of energy utilization, therefore, it is necessary for materials to regenerate a stable sorption capacity and improve the kinetic properties in the steps of both sorption and desorption.

(3) Representative significant research contents related to "CO_2 Capture" (2001).

Johnson and Keith [41] discussed that CCS technology has been regarded as an important technical measure to solve the contradiction between fossil-based electricity production and environmental issues related to climate changes, the adoption of CCS technology can contribute greatly to decreasing the economical expenditure of dealing with carbon dioxide emissions in electric sector. Huang et al. [42] described that there are two steps for current carbon storage technology, firstly, to use an amine to separate and capture the CO_2, and then, to pressurize the gas to supercritical carbon dioxide liquid. They found that the Dual Alkali Approach technique could be significant to enhance the efficiency in separating and capturing the carbon dioxide.

(4) Representative significant research contents related to "sequestration" (2000).

White et al. [43] reported that, as one of the most urgent and severe issues in the area of the environment, global warming has become a worldwide issue and need to be confronted and solved by every nation. They described that carbon storage refers to sequestrating the carbon dioxide in a designated location for a long term, therefore, it has become an effective technical means to mitigate the problem of global warming. Nowak and Crane [44] illustrated that CO_2 is an important part of greenhouse gases, a larger utilization of fossil-based energy production would be responsible for the emissions growth of CO_2, and urban forests can be tremendous useful for the decline of carbon dioxide in atmosphere, however, urban trees would also have negative effects on ecosystem. Stewart and Hessami [45] described that the sequestration of CO_2 was not yet mature, it is turned out that the techniques—no matter whether geologic or oceanic injectionare—still not sustainable, therefore, they present a possible technical measure to minimize the emissions of GHGs. Chu [46] reported that the energy production of fossil-based fuel leads to the growth of CO_2 emissions and causes severe climate change problems, and it is important to drastically decrease the CO_2 emissions in order to avoid possible environmental risks in the future. Herzog [47] examined the technical means related to carbon storage which involve processes of removing the GHGs and storing them in a selected reservoir. He described that, once a large quantity of carbon is captured, it would be ideal to achieve its commercial use, however, it also have some limitations in large-scale applications, therefore, as to a larger amount of captured carbon, it may be better to store it in geology or ocean.

(5) Representative significant research contents related to "solubility" (2007).

Suekane et al. [48] reported that, as to residual carbon dioxide stored in geology, solubility trapping is significant important in aquifer storage, some of the CO_2 stored in geology will dissolve and mix in the geological formation water. As to carbon storage technology, solubility trapping and residual gas can help improve storage capacity and efficiency. Keppler et al. [49] described that carbon solubility in olivine will be useful for capturing the dynamics of carbon dioxide exchange. They found that carbon solubility in olivine is unexpected low, which is fundamentally different from previous studies. Mitchell et al. [50] investigated the potential of microorganisms for improving the capacity and efficiency of CCS technology by using the techniques of solubility trapping and mineral-trapping.

(6) Representative significant research contents related to "performance" (2006).

Rubin and colleagues [51] reported that technological performance and economic expenditure are significant influencing factors in policy analysis of CCS technology. They found that, the emissions of carbon dioxide produced by power plant would be greatly reduced by CCS technology. Davision [52] evaluated the performance and costs of three main technologies for carbon capture employed in power plants, and found that, the economical expenditure of electricity would be expected to reduce by 18% under their designed technical conditions, and the advances in carbon dioxide capture technology would contribute to the reduction of economical expenditure. Abu-Zahra et al. [53] evaluated that to deploy CCS technology in a generating station will increase the expenditure of its electricity production, and the improvement and optimization of process will contribute to reducing the entire expenditure of carbon capture technology.

(7) Representative significant research contents related to "metal organic framework" (2009).

Sumida et al. [54] reported that CCS technology can efficiently capture carbon dioxide emitted from industrial sources, however, it can play an important role until the energy infrastructure has been modified greatly. They analyzed that, in order to decrease the cost related to CCS technology, it is essential to pay attention to the influencing factors related to the performance of carbon capture technology. And then, they described that metal-organic frameworks can be employed in carbon capture technology as an emerging and ideal materials. Millward and Yaghi [55] also analyzed several benefits of metal-organic frameworks for carbon storage technology. Li et al. [56] reported that, taking cost into account, to identify the emerging materials will become a significant issue in the research area of CCS technology. They examined that, because metal-organic frameworks have advantages in the cost and efficiency of synthesis, they can be greatly developed in the area of carbon capture technology as a separation material.

(8) Representative significant research contents related to "Monoethanolamine" (2007).

Freguia and Rochelle [57] reported that carbon storage technology has become an important technical means to cope with the environmental problem of global warming. They also described that, as an effective solvent for carbon capture, absorption with aqueous monoethanolamine has achieved practical application status. Strazisar and White [58] described that, in the technical process of CCS, how to separate carbon dioxide is one of the most critical parts, besides, they also analyzed that, as an effective and mature technology, monoethanolamine-based chemical absorption techniques have been widely practiced in the industry.

3.3. Analysis of Evolution and Emerging Trends

3.3.1. Burst Analysis and Research Evolution

Burstness value changes over time reflect when an abrupt change of frequency occurs in a specific duration [59]. In other words, an entity with a frequency burst value implies that its frequency has suddenly changed during a period of time [60]. As one of burst detection methods, citation burst can be considered as an important criterion to identify the most active area in a certain research field. A keyword node with a strong citation burst would represent the emerging trend of a specific research, and both the citation burst and occurrence burst can be supported by CiteSpace [61]. The burst means a major variable change in a given time period. This type of change is considered as an effective method to detect the emerging trends or academic frontiers in a particular research domain [62]. Several types of nodes, such as author, institutions, and keywords can be analyzed by burst detection. If applied for the node type of keywords, it will illustrate the fast-growing topics in the field being studied [61]. Therefore, the analysis of burst keywords is employed to investigate the emerging trends and their changes over the time in the research area of CCS technology. The top 300 keywords, whose citation burst value are relatively stronger, were selected for the analysis.

To identify the general situation of research trends and frontiers of CCS technology, the number distribution of burst terms in different years is analyzed and shown in Figure 6. It can be seen from the trend line of the polynomial equation that the annual number of burst keywords in the research field of CCS technology is somewhat similar to a "Smiling Curve". The evolution as a whole can be divided into three stages: A decline before 2005, stable growth between 2006 and 2013, and a rapid rise after 2014. It can also be estimated that the number of burst keywords will continue to increase over the next two years.

There are some severe challenges on issues of environmental protection and climate change in this 21st century [63]. As an effective technical means, CCS technology has attracted a great deal of attention from both academic and industrial fields, and it has become one of the most important research trends in the discipline of environment at the dawn of the new millennium. Especially, after an investigation on CCS technology was released by the IPCC in 2005, endeavours from relevant international organizations and countries also caused CCS technology to receive great attention during the 2nd stage. However, at this stage, academic research was also affected by some realistic factors, such as immaturity of the technology, technological risks, and the expensive costs. At present, as demonstration projects of CCS technology turn into practice, the relevant research has to be carried out in a comprehensive way. It can be concluded that there will be more branches and new developments in the research field of CCS technology in the next few years.

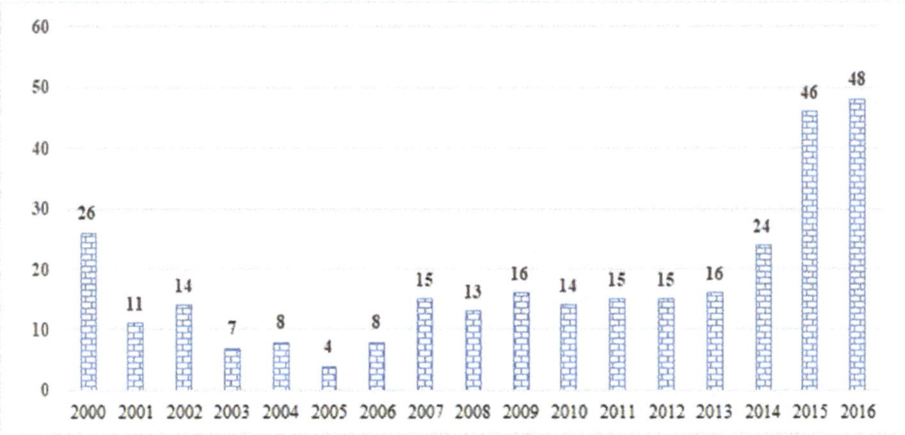

Figure 6. Annual distribution of TOP 300 burst keywords.

The citation burst of research literature represents when particular attention is paid to a specific domain of technology. The emerging trends in a particular research area can be identified by the analysis of literature with a stronger citation burst [64]. There are many pieces of literature with citation bursts and the top 20 pieces of literature are listed in Table 5. These research studies not only have a stronger burst strength, but also have active impact (the burst duration is lasting until the present) on the research area of CCS technology. Among the top 20 pieces of literature, the articles written by Boot-Handford [65] and published in *Energy & Environmental Science* ranked at the first place on the list with a burst strength of 30.447, in which the feasibility and technologies of carbon capture were reviewed, while the second strongest citation burst was written by Wang [66] and that is also established in *Energy & Environmental Science*. This paper felt that it is essential to update the earlier studies of carbon capture in a timely fashion, so it described how to organize the published *Nature* with a burst strength of 17.12; this paper provided a mechanistic framework, which is designed as a highly efficient adsorbents for removing the CO_2 [67]. In addition, there is also one reference published in *Science* with a burst strength of 10.3454. In that study, Furukawa et al. [68] focused on several aspects of metal-organic framework (MOF) chemistry, including its key developments and the impacts on practice. It can be considered that the literature studies listed in the table are the most influential works in the research area of CCS technology in recent years. They sorbents according to different working temperatures. There is one reference published in *Nature* with a burst strength of 17.12; this paper provided a mechanistic framework, which is designed as a highly efficient adsorbents for removing the CO_2 [67]. In addition, there is also one reference published in *Science* with a burst strength of 10.3454. In that study, Furukawa et al. [68] focused on several aspects of metal-organic framework (MOF) chemistry, including its key developments and the impacts on practice. It can be considered that the literature studies listed in the table are the most influential works in the research area of CCS technology in recent years. They can represent the recent advances of CCS technology and meanwhile, the topics of these pieces of literature can represent the emerging trends and latest developments of CCS technology.

Table 5. Top 20 literature with strongest citation bursts.

Reference Information	Year	Strength	Begin	End
Boot-Handford ME, 2014, Energ Environ Sci, V7, P130	2014	30.447	2015	2018
Wang JY, 2014, Energ Environ Sci, V7, P3478	2014	29.6231	2016	2018
Leung DYC, 2014, Renew Sust Energ Rev, V39, P426	2014	22.3315	2016	2018
Mcdonald TM, 2015, Nature, V519, P303	2015	17.12	2016	2018
Zhang ZJ, 2014, Energ Environ Sci, V7, P2868	2014	16.1407	2015	2018
Dutcher B, 2015, Acs Appl Mater Inter, V7, P2137	2015	13.9032	2016	2018
Chowdhury FA, 2013, Ind Eng Chem Res, V52, P8323	2013	13.2412	2016	2018
Kenarsari SD, 2013, Rsc Adv, V3, P22739	2013	10.7725	2015	2018
Thommes M, 2015, Pure Appl Chem, V87, P1051	2015	10.4521	2016	2018
Li YJ, 2015, Appl Energ, V145, P60	2015	10.4521	2016	2018
Furukawa H, 2013, Science, V341, P974	2013	10.3454	2015	2018
Shekhah O, 2014, Nat Commun, V5, P	2014	10.1986	2015	2018
Yu Ch, 2012, Aerosol Air Qual Res, V12, P745	2012	9.281	2016	2018
Goto K, 2013, Appl Energ, V111, P710	2013	9.0674	2016	2018
IPCC, 2014, Climate Change 2014, V, P	2014	9.0578	2016	2018
Sethia G, 2015, Carbon, V93, P68	2015	9.0578	2016	2018
Dai Zd, 2016, J Membrane Sci, V497, P1	2016	9.0578	2016	2018
Liang Zw, 2015, Int J Greenh Gas Con, V40, P26	2015	9.0578	2016	2018
To Jwf, 2016, J Am Chem Soc, V138, P1001	2016	8.7093	2016	2018
Fracaroli Am, 2014, J Am Chem Soc, V136, P8863	2014	8.5977	2015	2018

Table 6 lists the top 10 keywords with stronger citation bursts in different periods of time, the research topics and their changes in the research area of CCS technology are presented in the table. In order to investigate the burst strength of different keywords, keywords with the same meaning, but expressed by different words/phrases, for example carbon dioxide and CO_2, would not be merged. The table shows the first time that each keyword appeared and its duration. What should be noted is that the line illustrated by blue color in the table means the whole study period of time (2000–2018) and the red line represents the duration of the citation burst. In addition, in order to understand more exactly the emerging trends in the research area of CCS technology and their changes over time, the research period is divided into three parts.

It can be seen that the top 10 burst keywords in different time periods were essentially different. During the time period from 2000 to 2005, "carbon dioxide" is at the first place in the list with a burst strength value of 52.5794 and the second keyword was "CO_2" with a burst strength of 45.0923. During the time period from 2006 to 2010, "high capacity" had the strongest citation burst with a strength of 15.9928, followed by "Monte Carlo simulation" (14.5793), and "calcium oxide" (11.5709). Furthermore, during the time period of from 2011 to present, "methane" had the strongest citation burst with a strength of 10.672, followed by "post-combustion CO_2 capture" (9.9749) and "carbide slag" (7.6256). The keywords listed in the table reflect the CCS research characteristics, such as hotspots, emerging trends and new developments in different periods.

Table 6 also shows the different burst durations of each keyword in different periods of time. This duration reflects the length of keywords' influence in the research area of CCS technology. As is shown, most of the keywords in the research period of 2000–2005, such as "dynamics", "CO_2", "hydrogen", "carbon storage", "forest", "bioma", and "disposal" had a longer burst duration than the keywords in the second and the third research periods. It is safe to assume that the studies related to these keywords can be regarded as the research pioneers in the domain of CCS technology. They can serve as important and fundamental sources for technological development and new research branches, so that the research studies embodied in these keywords have a longer influence on the advancements in this technological domain.

Table 6. Top 10 keywords with strong citation bursts in different periods.

Period	Top 10 Keywords	Strength	Begin	End	2000–2018
2000–2005	Carbon dioxide	52.5794	2001	2008	
	CO_2	45.0923	2001	2010	
	Nitrogen	21.3001	2000	2008	
	Hydrogen	20.7827	2003	2012	
	Ecosystem	20.6307	2000	2008	
	Carbon storage	20.2215	2000	2009	
	Forest	19.5339	2000	2009	
	Bioma	17.2279	2000	2009	
	Disposal	16.4679	2002	2011	
	Dynamics	16.4412	2000	2010	
2006–2010	High capacity	15.9928	2009	2013	
	Monte Carlo simulation	14.5793	2006	2011	
	Calcium oxide	11.5709	2006	2010	
	CaO	11.5081	2007	2010	
	Gas	11.0606	2006	2010	
	Combustion	10.2953	2008	2010	
	Calcination	9.9176	2009	2010	
	SBA 15	9.832	2007	2011	
	Reactor	9.4293	2007	2011	
	Carbonation	9.2963	2008	2012	
2011-present	Methane	10.672	2011	2012	
	Post-combustion CO_2 capture	9.9749	2016	2018	
	Carbide Slag	7.6256	2016	2018	
	Gas Storage	7.5231	2014	2016	
	Fabrication	7.2958	2016	2018	
	Graphene	7.252	2015	2018	
	Ambient Condition	6.899	2016	2018	
	Sulfation	6.8989	2012	2013	
	Facile Synthesis	6.701	2016	2018	
	Expanded Mesoporous Silica	6.4517	2011	2013	

Furthermore, the keywords in the research period of 2011-present, such as "post-combustion CO_2 capture", "carbide slag", "fabrication", "graphene", "ambient condition" and "facile synthesis", can represent the new topics, emerging trends, and even the major frontiers in the research area of CCS technology, because their burst durations have lasted until the present. The studies related to these keywords may have a continuous impact on the development of CCS technology.

3.3.2. Main Contents of Important Emerging Trends

(1) Representative significant research contents related to "post-combustion CO_2 capture" (burst strength: 9.9749; burst duration: 2016–2018).

Abanades et al. [69] described that CCS technology can be considered as one of the significant technical means to cope with climate change, and it is necessary to analyze the economical expenditure to explore the potential paths of utilizing carbon capture technology. They analyzed that the most common form of the cost estimation of carbon capture is incremental expenditure in power and the expenditure in removing the CO_2. Veltman and colleagues [71] evaluated the impacts of carbon capture technology on the environment and human health. It is indicated that if the scrubbing techniques based on amines are adopted in carbon capture technology, the toxicity to the freshwater ecosystem will increase 10 times, and the toxicity of terrestrial ecosystems will also suffer a minor increase. Plaza et al. [71] made a study on two different carbon adsorbent approaches and found that both of the techniques can be used to produce the adsorbents with a better CO_2 adsorption efficiency, meanwhile, they also considered that nitrogen functionalities can contribute to enhancing the carbon dioxide adsorption efficiency, especially under the condition of low partial pressures.

(2) Representative significant research contents related to "carbide slag" (burst strength: 7.6256; burst duration: 2016–2018).

Li et al. [72] introduced the sources and properties of carbide slag, and then, they investigated a new synthetic sorbent by using the combustion synthetic techniques and variety of materials, including carbide slag, and found that porous structures can be used to increase the capabilities of synthetic sorbenta, moreover, the capabilities of synthetic sorbenta will be beneficial to capture carbon dioxide. Li et al. [73] investigated that, in order to decrease and eliminate the carbon dioxide, the calcium looping technology is regarded as a valuable and feasible techniques for capturing carbon dioxide, and then, they concluded that, comparing with limestone or Hy-CaO, carbide slag have advantages in final carbonation conversion under the same technical conditions. Sun et al. [74] reported that, integration of a carbide slag disposal and a calcium looping system can contribute to decreasing the expenditure of transporting the raw sorbent materials. Thus, they described that the pelletization of carbide slag is very important for calcium looping process, in order to avoid appreciable loss of CO_2 sorption, it is also essential to control the calcination temperature.

(3) Representative significant research contents related to "fabrication" (burst strength: 7.2958; burst duration: 2016–2018).

Yu and colleagues [75] reported that, although amine baths are the commonly used techniques for CCS technology in practical application, the process itself also has some obvious disadvantages. Thus, in order to overcome the disadvantages, they fabricated bimodal mesoporous silica hollow spheres and considered that the CO_2 adsorption efficiency may depend on the temperature of adsorption and the specific surface areas of the sample. Qin et al. [76] analyzed that the process of carbon capture contributes about 75% of the total cost in CCS technology, however, in fact, only a few of the proposed technologies can be technically feasible and cost-effective in industrial practice, and currently, amine scrubbing is the only technique that is relatively mature and has commercial application possibilities. Therefore, they intended to find an efficient and low-cost fabricating and sorbent technology using the method of extending wet mixing and a cheaper insoluble precursors. Sun et al. [77] reported that, at present, the separation of carbon dioxide from N_2 and CH_4 has attracted increasing attention, the common method to eliminate carbon dioxide is "wet scrubbing," however, this process itself has several inherent shortcomings. Therefore, they designed a particular reaction to fabricate porous polymer networks. Liu et al. [78] presented a versatile fabrication strategy to conquer the defects of hierarchically porous carbon materials in industrial application.

(4) Representative significant research contents related to "graphene" (burst strength: 7.252; burst duration: 2015–2018).

Chandra et al. [79] reported two commonly used technologies and their technological limitations, and then, they considered that, as a promising and emerging technology, graphene can have several obvious advantages in the area of carbon absorption technology. Therefore, they examine how to synthesize one kind of porous carbon using graphene. Garcia Gallastegui et al. [80] reported that graphene itself has several very good technical properties, so graphene is receiving intense attention, and has been explored in various application fields. Besides, layered double hydroxides (LDHs) are also useful materials for carbon dioxide adsorbents. Therefore, in their study, they described that graphene oxide is helpful for improving the adsorption efficiency of carbon dioxide of LDH.

Li et al. [81] reported that graphene oxide is a good nanofiller in mixed matrix membranes (MMMs). Thus, they developed a novel multi-permselective MMMs in their experiments and found that the polyethylenimine-functionalized graphene oxide nanosheets has a positive role in promoting membrane properties. Shen et al. [82] reported that graphene oxide is a promising technology in membranes for molecular separation, and then, they proposed the membranes with carbon dioxide transportation channels of graphene oxide laminates.

(5) Representative significant research contents related to "facile synthesis" (burst strength: 6.701; burst duration: 2016–2018).

Li et al. [83] described a synthesis of hierarchical mesoporous carbon nitride spheres, and then, they investigated the carbon dioxide sorption performance of the products and found that, because of the capillary condensation effect, the micropores and mesopores can improve the capture properties of carbon dioxide. Lu and Zhang [84] reported that, because nanoporous organic polymers has many very good technical properties, they are regarded as a valuable and emerging material in the area of CCS technology, however, it is important for improving the adsorption efficiency to determine the porosity parameters of porous polymers. Kim and colleagues [85] reported that, the adsorbents for capturing the carbon at a higher temperature have attracted lots of attention in recent years. Thus, they tried to synthesize macroporous Li_4SiO_4 by using the method of solid-state transformation, which can significantly enhance carbon dioxide adsorption performance compared with conventional Li_4SiO_4.

4. Discussion

Knowledge mapping can provide us with valuable information about the status and trends of a certain field. It is turned out that CiteSpace is a useful tool for knowledge mapping. Literature mining and analysis is not only useful for researchers to find new research directions, but also benefits potential inventors/governmental agencies who intend to invest in mature technologies [86]. In this study, a comprehensive investigation about the research evolution and developing trends of CCS technology is provided using the CiteSpace software, based on retrieving the research literature from Web of Science. Some valuable conclusions can be illustrated from the perspectives of basic analysis, research topics analysis, and emerging trends and new development.

At present, there are growing concerns about reducing CO_2 emissions, so it is necessary to adopt emerging technologies in addition to the utilization of low carbon and renewable energy sources [87]. Among many possible technical means to cope with global warming, CCS technology is regarded as an effective technical means for removing carbon dioxide [88,89]. CCS technology is an integrated suite of technologies which can capture and store CO_2 at a selected place and prevent CO_2 from being emitted directly to the atmosphere, so that, many endeavors for the promotion of CCS technology have been initiated recently [90]. According to the report released by the IPCC in 2005, over 60% of global carbon dioxide emissions come from point sources that could apply carbon capture technology. CCS technology can contribute up to 20% of needed carbon dioxide emission reductions over the coming century [31]. In 2007, the United Nations Climate Conference was held and an agreement on the commitment of countries' carbon reduction and responsibility was promoted through a meeting of the Bali Road Map [91]. In December 2009, the Copenhagen Accord was drafted to reaffirm the goal of carbon reduction. The global CCS patent applications increased rapidly after 2006, the organizations from the U.S., China and some European countries, filed more patents [92]. Therefore, as one of promising technologies to combat climate change, CCS technology has received extensive attention, and the technological innovation activities in this particular technological domain have been promoted greatly since 2005.

In the CCS research area, there are more studies published in countries from Asia, North America, and Europe, especially in China and the U.S. These two countries are not only participating countries of the Copenhagen Accord, but also the main two sponsors of the accord. In the United States, carbon capture and storage technology has been regarded as an important consideration in U.S. climate policy discussions, and the funding for the research and demonstration of CCS technology has experienced a sharp increase in the new century [93]. In China, as the world's biggest emitter of GHGs and consumer of energy production, the government has provided active support, including finances and policies, for promoting the development of CCS technology in China [94]. In 2015, the advancements in CCS technology has even been proposed in National Medium and Long Term Development Plan of Science and Technology of China [95]. Moreover, in Japan, a long-term research program was formulated for CCS technology. In Europe, the first international conference on CCS technology was held in 1992 in Amsterdam [47]. According to an investigation made in 2006 about CCS technology from 28 European countries, 75% of the European energy stakeholders thought that it was "definitely" or "probably necessary" to widely deploy CCS technology to reduce the CO_2 emissions in their own countries [96]. Moreover, in order to deal with the challenges of sustainable energy and external energy dependency, the European Union has also developed some legislative initiatives, such as "European 2020", "Energy Roadmap 2050", and so on [97].

5. Conclusions

As CCS technology has a significant role in controlling CO_2 emissions, many countries have started to deploy CCS technology. With the vigorous support and push of governments, a multitude of research studies were carried out and the number of research articles increased in these countries correspondingly. In addition, the most productive institutions in the research area of CCS technology were identified in this study. There are several institutions from China. The CAS of China not only has the largest number of publications but also has the largest number of cited frequency and centrality. However, among all the most productive institutions, more institutes from developed countries have significant roles and greater impact on CCS research. In addition, among the main institutions, they linked closely with each other, which means that most of them paid more attention to cooperating and exchanging with each other in the scientific research of CCS technology.

In the research area of CCS technology, "carbon dioxide", "CO_2 capture", "adsorption", "separation", and "adsorption" are the major research topics while "hydrogen storage", "membrane", "polymer", and "carbon sequestration" are the research branches. "CO_2", "Kinetics", "CO_2 capture", and "sequestration" are the core research contents and "carbon dioxide capture" is the important research foundation while "performance", "metal organic framework", "technology", "solubility", and "monoethanolamine" are the new research topics. The important pieces of literature related to the research topics mentioned above and their citation network were also investigated. The significant findings of the research topics were summarized by reviewing the most cited literature related to the keywords with higher value of "centrality". The contents of the main research topics in the CCS technology include at least the following aspects: (1) Carbon emissions and environmental protection. The impact of carbon emissions on environment is addressed in the whole research period [33,34,39,43], it could be concluded that this topic would be continuous. Therefore, in order to cope with the severe problems posed by climate change and global warming, it is necessary for us to develop green and efficient technologies [35,36]. (2) Research and development of CCS technology. This topic focuses on several issues, such as experiment of carbon adsorption or sorption [36,37,40,46], new process for carbon capture or storage [42], porous materials (metal organic framework or metal organic materials) for carbon adsorption or separation [30,54], applications of monoethanolamine in carbon capture [57,58], applications of solubility trapping in carbon storage technologies [48–50], and so on. (3) Social practical issues of CCS technology. This topic reflects the degree of development or feasibility of CCS technology in practice. This topic includes not only the assessment of technological capabilities [48,50], but also the cost analysis and economic performance of CCS technology [41,51–53].

The keywords, such as "carbon dioxide", "CO_2", "high capacity", and "methane" represented emerging trends of CCS research areas in different periods of time and the keywords, such as "post-combustion CO_2 capture", "carbide slag", "fabrication", "graphene", "ambient condition", and "facile synthesis" may be new developments and possibly the major frontiers of the emerging trends in the research area of CCS technology. Similarly, important literature studies related to emerging trends were also stated. The valuable information about the latest advances and developments of the research frontiers of CCS technology has been analyzed using the burst analysis of keywords and literature. Specially, the important findings of the emerging trends were summarized by reviewing the most cited literature related to the keywords whose burst duration last until the present (2018). The contents of the main emerging trends in the CCS technology include at least the following aspects: (1) Emerging techniques and processes. With the development of science and technology, relevant techniques and processes have been developed and applied in CCS technology, such as postcombustion [69,70,98], combustion synthetic method [72], calcium looping technology [73,74], amine-functionalized solids and amine-scrubbing technique [75,76], facile nucleophilic substitution reaction [78], and so on. (2) Emerging materials. In order to increase technological efficiency, some new materials, including graphene [79–82,99] and nanoporous organic polymers [84], are employed in the innovative activities of CCS technology. (3) Evaluation of technological performance. Technological performance corresponds to the capabilities of firms to exploit the potentialities of the prevailing technology, and it can evaluated by the productivity level of firms [100]. In recent years, in the research area of CCS technology, technological performance is assessed from several aspects, such as performance of carbon capture [83], adsorption capability of carbon [71,75,80,85], performance of carbon sorbent [72] and performance of carbon membranes [81,82], and so on. (4) Socioeconomic analysis. This research trend is related to cost of carbon capture [70], cost of carbon transportation [74], economic performance of CCS technology [76,101], human health and environment [70,102].

6. Research Limitations

Generally speaking, the knowledge mapping of CCS technology is explored in this study. It can provide researchers or practitioners with explicit information about the research situations and emerging trends of CCS technology. With the continuous development of CCS technology, the research topics in this particular research area also changing. The main contents of research topics in the field of CCS technology involve at least following three aspects: carbon emissions and environmental protection, research and development (R&D) activities, and social practical issues. Additionally, as CCS technology tends to be gradually mature and its important role in mitigating climate change is being realized, which not only promotes the further development of CCS technology, but also inspired some emerging trends in the research area, such as emerging techniques and processes, emerging materials, evaluation of technological performance, and socioeconomic analysis, and more. However, it also should be noted that there are still some limitations in this study. First of all, the studies were collected from the Web of Science. As is known, the Web of Science is one of the most important bibliographic database in the world. Certainly, the dataset used in this study can represent the situation, hotspots, trends, and frontiers of the research area of CCS technology, but not all publications/journals are indexed by the database, especially in non-English speaking countries, and this case would be especially true. There are always some important publications published in the languages of these countries. Therefore, these publications would be excluded by this study. In addition, when it comes to CiteSpace, this software has some requirements about the data format, which would have an impact on the scope of data collection and analysis. These limitations should be carefully considered in any future relevant research.

Author Contributions: Hong-Hua Qiu designed this study, wrote the paper and performed all the research steps. Lu-Ge Liu searched and preprocessed the literature and ran the software of CiteSpace. All the authors cooperated, discussed and revised the paper.

Acknowledgments: This research is supported by the National Social Science Foundation of China under Grant no. 15BTQ047. The corresponding author is grateful for the financial sponsorship of Chinese Scholarship Council (CSC) to support his study in Max Planck Institute for Innovation and Competition as a visiting scholar. Finally, special thanks to my supervisor, Dietmar Harhoff, the director of Max Planck Institute for Innovation and Competition. He not only provides me with very good research conditions but also gives me many valuable comments for my research when I carry out my research at MPI.

Conflicts of Interest: The authors declare no conflicts of interest.

References

1. Quéré, C.L.; Raupach, M.R.; Canadell, J.G.; Marland, G.; Bopp, L.; Ciais, P.; Conway, T.J.; Doney, S.C.; Feely, R.A.; Friedlingstein, P.F.P.; et al. Trends in the sources and sinks of carbon dioxide. *Nat. Geosci.* **2009**, *2*, 831–836. [CrossRef]
2. Haszeldine, R.S. Carbon capture and storage: How green can black be? *Science* **2009**, *325*, 1647–1652. [CrossRef] [PubMed]
3. Benson, S.M.; Bennaceur, K.; Cook, P.; Davison, J.; de Coninck, H.; Farhat, K.; Ramirez, C.A.; Simbeck, D.; Surles, T.; Verma, P.; et al. Carbon Capture and Storage. In *Global Energy Assessment—Toward a Sustainable Future*; Cambridge University Press: Cambridge, MA, USA, 2012; pp. 993–1068, ISBN 9781 10700 5198.
4. IEA. *Technology Roadmap: Carbon Capture and Storage 2013*; International Energy Agency: Paris, France, 2013. Available online: http://www.iea.org/publications/freepublications/publication/TechnologyRoadmapCarbonCaptureandStorage.pdf (accessed on 2 February 2018).
5. Broecks, K.P.F.; Egmond, S.V.; Rijnsoever, F.J.V.; Berg, M.V.D.; Hekkert, M.P. Persuasiveness, importance and novelty of arguments about Carbon Capture and Storage. *Environ. Sci. Policy* **2016**, *59*, 58–66. [CrossRef]
6. Intergovernmental Panel on Climate Change, Carbon Dioxide Capture and Storage. Available online: http://www.ipcc.ch/pdf/special-reports/srccs/srccs_wholereport.pdf (accessed on 30 January 2018).
7. Kern, F.; Gaede, J.; Meadowcroft, J.; Watson, J. The political economy of carbon capture and storage: An analysis of two demonstration projects. *Technol. Forecast. Soc.* **2016**, *102*, 250–260. [CrossRef]
8. De Souza, E.F.; de Almeida Falbo, R.; Vijaykumar, N.L. Knowledge management initiatives in software testing: A. mapping study. *Inf. Softw. Technol.* **2015**, *57*, 378–391. [CrossRef]
9. Powell, W.W.; Snellman, K. The Knowledge Economy. *Annu. Rev. Sociol.* **2004**, *30*, 199–220. [CrossRef]
10. Lyu, H.; Zhou, Z.; Zhang, Z. Measuring Knowledge Management Performance in Organizations: An Integrative Framework of Balanced Scorecard and Fuzzy Evaluation. *Information* **2016**, *7*, 29. [CrossRef]
11. Darroch, J. Knowledge management, innovation and firm performance. *J. Knowl. Manag.* **2005**, *9*, 101–115. [CrossRef]
12. Centobelli, P.; Cerchione, R.; Esposito, E. Knowledge Management in Startups: Systematic Literature Review and Future Research Agenda. *Sustainability* **2017**, *9*, 361. [CrossRef]
13. Hall, R.; Andriani, P. Managing Knowledge for Innovation. *Long Range Plan.* **2002**, *35*, 29–48. [CrossRef]
14. Ebener, S.; Khan, A.; Shademani, R.; Compernolle, L.; Beltran, M.; Lansang, M.; Lippman, M. Knowledge mapping as a technique to support knowledge translation. *Bull. World Health Organ.* **2006**, *84*, 636–642. [CrossRef] [PubMed]
15. Fisher, K.M. *Overviee of Knowledge Mapping//Mapping Biology Knowledge*; Springer: Dordrecht, The Netherlands, 2002; pp. 5–23. Available online: https://link.springer.com/content/pdf/10.1007/0-306-47225-2_2.pdf (accessed on 28 January 2018).
16. Vailiii, E. Knowledge Mapping: Getting Started with Knowledge Management. *J. Inf. Syst. Manag.* **1999**, *16*, 16–23. [CrossRef]
17. Mingers, J.; Leydesdorff, L. A review of theory and practice in scientometrics. *Eur. J. Oper. Res.* **2015**, *246*, 1–19. [CrossRef]
18. Mercuri, E.G.F.; Kumata, A.Y.J.; Amaral, E.B.; Vitule, J.R.S. Energy by Microbial Fuel Cells: Scientometric global synthesis and challenges. *Renew. Sustain. Energy Rev.* **2016**, *65*, 832–840. [CrossRef]
19. Montoya, F.G.; Montoya, M.G.; Gómez, J.; Manzano-Agugliaro, F.; Alameda-Hernández, E. The research on energy in Spain: A scientometric approach. *Renew. Sustain. Energy Rev.* **2014**, *29*, 173–183. [CrossRef]
20. Konur, O. The scientometric evaluation of the research on the algae and bio-energy. *Appl. Energy* **2011**, *88*, 3532–3540. [CrossRef]
21. Staab, S.; Studer, R.; Schnurr, H.P.; Sure, Y. Knowledge Processes and Ontologies. *IEEE Intell. Syst.* **2001**, *16*, 26–34. [CrossRef]
22. Chen, C. CiteSpace II: Detecting and visualizing emerging trends and transient patterns in scientific literature. *J. Assoc. Inf. Sci. Technol.* **2006**, *57*, 359–377. [CrossRef]
23. Chen, C. Searching for intellectual turning points: Progressive knowledge domain visualization. *Proc. Natl. Acad. Sci. USA* **2004**, *101* (Suppl. 1), 5303–5310. [CrossRef] [PubMed]

24. Yu, D.; Xu, C. Mapping research on carbon emissions trading: A co-citation analysis. *Renew. Sustain. Energy Rev.* **2017**, *74*, 1314–1322. [CrossRef]

25. Chen, K.; Guan, J. A bibliometric investigation of research performance in emerging nanobiopharmaceuticals. *J. Informetr.* **2011**, *5*, 233–247. [CrossRef]

26. Lin, Z.; Wu, C.; Hong, W. Visualization analysis of ecological assets/values research by knowledge mapping. *Acta Ecol. Sin.* **2015**, *35*, 142–154. [CrossRef]

27. D'Alessandro, D.M.; Smit, B.; Long, J.R. Carbon dioxide capture: Prospects for new materials. *Angew. Chem. Int. Ed.* **2010**, *49*, 6058–6082. [CrossRef] [PubMed]

28. Choi, S.; Drese, J.H.; Jones, C.W. Adsorbent materials for carbon dioxide capture from large anthropogenic point sources. *ChemSusChem* **2009**, *2*, 796–854. [CrossRef] [PubMed]

29. Rochelle, G.T. Amine scrubbing for CO_2 capture. *Science* **2009**, *325*, 1652–1654. [CrossRef] [PubMed]

30. Nugent, P.; Belmabkhout, Y.; Burd, S.D.; Cairns, A.J.; Luebke, R.; Forrest, K.; Pham, T.; Ma, S.; Space, B.; Wojtas, L.; et al. Porous materials with optimal adsorption thermodynamics and kinetics for CO_2 separation. *Nature* **2013**, *495*, 80–84. [CrossRef] [PubMed]

31. Metz, B.; Davidson, O.; de Coninck, H.C.; Loos, M.A.; Meyer, L.A. *IPCC Special Report on Carbon Dioxide Capture and Storage*; Cambridge University Press: New York, NY, USA, 2005; ISBN 0-521-68551-6.

32. Xiang, C.; Wang, Y.; Liu, H. A scientometrics review on nonpoint source pollution research. *Ecol. Eng.* **2017**, *99*, 400–408. [CrossRef]

33. Hobbie, S.E.; Schimel, J.P.; Trumbore, S.E.; Randerson, J.R. Controls over carbon storage and turnover in high-latitude soils. *Glob. Chang. Biol.* **2000**, *6*, 196–210. [CrossRef]

34. Schimel, D.; Melillo, J.; Tian, H.; McGuire, A.D.; Kicklighter, D.; Kittel, T.; Rosenbloom, N.; Running, S.; Thornton, P.; Ojima, D. Contribution of increasing CO_2 and climate to carbon storage by ecosystems in the United States. *Science* **2000**, *287*, 2004–2006. [CrossRef] [PubMed]

35. Treseder, K.K.; Allen, M.F. Mycorrhizal fungi have a potential role in soil carbon storage under elevated CO_2 and nitrogen deposition. *New Phytol.* **2000**, *147*, 189–200. [CrossRef]

36. Ravikovitch, P.I.; Vishnyakov, A.; Ron Russo, A.; Neimark, A.V. Unified Approach to Pore Size Characterization of Microporous Carbonaceous Materials from N2, Ar, and CO_2 Adsorption Isotherms. *Langmuir* **2000**, *16*, 2311–2320. [CrossRef]

37. Ding, Y.; Alpay, E. Equilibria and kinetics of CO_2, adsorption on hydrotalcite adsorbent. *Chem. Eng. Sci.* **2000**, *55*, 3461–3474. [CrossRef]

38. Sobkowski, J.; Czerwiński, A. Kinetics of carbon dioxide adsorption on a platinum electrode. *J. Electroanal. Chem. Interface* **1974**, *55*, 391–397. [CrossRef]

39. Serna-Guerrero, R.; Sayari, A. Modeling adsorption of CO_2, on amine-functionalized mesoporous silica. 2: Kinetics and breakthrough curves. *Chem. Eng. J.* **2010**, *161*, 182–190. [CrossRef]

40. Ochao-Fernandez, E.; Roenning, M.; Grande, T.; Chen, D. Nanocrystalline Lithium Zirconate with Improved Kinetics for High-Temperature CO_2, Capture. *Chem. Mater.* **2006**, *37*, 1383–1385. [CrossRef]

41. Johnson, T.L.; Keith, D.W. Electricity from fossil fuels without CO_2 emissions: Assessing the costs of carbon dioxide capture and sequestration in U.S. electricity markets. *J. Air Waste Manag. Assoc.* **2001**, *51*, 1452–1459. [CrossRef] [PubMed]

42. Huang, H.P.; Shi, Y.; Li, W.; Chang, S.G. Dual Alkali Approaches for the Capture and Separation of CO_2. *Energy Fuels* **2001**, *15*, 263–268. [CrossRef]

43. White, C.M.; Strazisar, B.R.; Granite, E.J.; Hoffman, J.S.; Pennline, H.W. Separation and Capture of CO_2 from Large Stationary Sources and Sequestration in Geological Formations-Coalbeds and Deep Saline Aquifers. *J. Air Waste Manag. Assoc.* **2003**, *53*, 645–715. [CrossRef] [PubMed]

44. Nowak, D.J.; Crane, D.E. Carbon storage and sequestration by urban trees in the USA. *Environ. Pollut.* **2002**, *116*, 381–389. [CrossRef]

45. Stewart, C.; Hessami, M.A. A study of methods of carbon dioxide capture and sequestration—The sustainability of a photosynthetic bioreactor approach. *Energy Convers. Manag.* **2005**, *46*, 403–420. [CrossRef]

46. Chu, S. Carbon capture and sequestration. *Science* **2009**, *325*, 1599. [CrossRef] [PubMed]

47. Herzog, H.J. What future for carbon capture and sequestration? *Environ. Sci. Technol.* **2001**, *35*, 148A–153A. [CrossRef] [PubMed]

48. Suekane, T.; Nobuso, T.; Hirai, S.; Kiyota, M. Geological storage of carbon dioxide by residual gas and solubility trapping. *Int. J. Greenh. Gas Control* **2008**, *2*, 58–64. [CrossRef]

49. Keppler, H.; Wiedenbeck, M.; Shcheka, S.S. Carbon solubility in olivine and the mode of carbon storage in the Earth's mantle. *Nature* **2003**, *424*, 414–416. [CrossRef] [PubMed]

50. Mitchell, A.C.; Dideriksen, K.; Spangler, L.H.; Cunningham, A.B.; Gerlach, R. Microbially enhanced carbon capture and storage by mineral-trapping and solubility-trapping. *Environ. Sci. Technol.* **2010**, *44*, 5270–5276. [CrossRef] [PubMed]

51. Rubin, E.S.; Chen, C.; Rao, A.B. Cost and performance of fossil fuel power plants with CO_2, capture and storage. *Energy Policy* **2007**, *35*, 4444–4454. [CrossRef]

52. Davison, J. Performance and costs of power plants with capture and storage of CO_2. *Energy* **2007**, *32*, 1163–1176. [CrossRef]

53. Abu-Zahra, M.R.M.; Niederer, J.P.M.; Feron, P.H.M.; Versteeg, G.F. CO_2, capture from power plants: Part II. A parametric study of the economical performance based on mono-ethanolamine. *Int. J. Greenh. Gas Control* **2007**, *1*, 135–142. [CrossRef]

54. Sumida, K.; Rogow, D.L.; Mason, J.A.; McDonald, T.M.; Bloch, E.D.; Herm, Z.R.; Bae, T.H.; Long, J.R. Carbon Dioxide Capture in Metal–Organic Frameworks. *Chem. Rev.* **2012**, *112*, 724–781. [CrossRef] [PubMed]

55. Millward, A.R.; Yaghi, O.M. Metal-organic frameworks with exceptionally high capacity for storage of carbon dioxide at room temperature. *J. Am. Chem. Soc.* **2005**, *127*, 17998–17999. [CrossRef] [PubMed]

56. Li, J.R.; Ma, Y.; Mccarthy, M.C.; Sculley, J.; Yu, J.; Jeong, H.; Balbuena, P.B.; Zhou, H. Carbon dioxide capture-related gas adsorption and separation in metal-organic frameworks. *Coord. Chem. Rev.* **2011**, *255*, 1791–1823. [CrossRef]

57. Freguia, S.; Gary, T. Rochelle. Modeling of CO_2 capture by aqueous monoethanolamine. *AIChE J.* **2003**, *49*, 1676–1686. [CrossRef]

58. Strazisar, B.R.; Anderson, R.R.; White, C.M. Degradation Pathways for Monoethanolamine in a CO_2 Capture Facility. *Energy Fuels* **2003**, *17*, 1034–1039. [CrossRef]

59. Kleinberg, J. Bursty and Hierarchical Structure in Streams. *Data Min. Knowl. Discov.* **2003**, *7*, 373–397. [CrossRef]

60. Chen, C.; Dubin, R.; Kim, M.C. Emerging trends and new developments in regenerative medicine: A scientometric update (2000–2014). *Expert Opin. Biol. Ther.* **2014**, *14*, 1295–1317. [CrossRef] [PubMed]

61. Chen, C. The CiteSpace Manual (Version 1.01). Available online: https://www.researchgate.net/profile/Ar sev_Aydinoglu/publication/274377526_Collaborative_interdisciplinary_astrobiology_research_a_biblio metric_study_of_the_NASA_Astrobiology_Institute/links/5670463b08ae0d8b0cc0e112.pdf (accessed on 18 February 2018).

62. Li, X.; Ma, E.; Qu, H. Knowledge mapping of hospitality research-A visual analysis using CiteSpace. *Int. J. Hosp. Manag.* **2017**, *60*, 77–93. [CrossRef]

63. Kojima, T. *Carbon Dioxide Problem: Integrated Energy and Environmental Policies for the 21st Century*; Gordon and Breach Science Publishers: Tokyo, Japan, 1998; ISBN 90-5699-127-2.

64. Yu, D.; Xu, Z.; Pedrycz, W.; Wanru, W. Information Sciences 1968–2016: A retrospective analysis with text mining and bibliometric. *Inf. Sci.* **2017**, *418*, 619–634. [CrossRef]

65. Boothandford, M.E.; Abanades, J.C.; Anthony, E.J.; Blunt, M.J.; Brandani, S.; Dowell, N.M.; Fernández, J.R.; Ferrari, M.; Gross, R.; Hallett, J.P.; et al. Carbon capture and storage update. *Energy Environ. Sci.* **2014**, *7*, 130–189. [CrossRef]

66. Wang, J.; Huang, L.; Yang, R.; Zhang, Z.; Wu, J.; Gao, Y.; Wang, Q.; O'Hare, D.; Zhong, Z.Y. Recent advances in solid sorbents for CO_2 capture and new development trends. *Energy Environ. Sci.* **2014**, *7*, 3478–3518. [CrossRef]

67. Mcdonald, T.M.; Mason, J.A.; Kong, X.; Bloch, E.D.; Gygi, D.; Dani, A.; Crocellà, V.; Giordanino, F.; Odoh, S.O.; Drisdell, W.S. Cooperative insertion of CO_2 in diamine-appended metal-organic frameworks. *Nature* **2015**, *519*, 303–308. [CrossRef] [PubMed]

68. Furukawa, H.; Cordova, K.E.; O'Keeffe, M.; Yaghi, O.M. The chemistry and applications of metal-organic frameworks. *Science* **2013**, *341*, 974. [CrossRef] [PubMed]

69. Abanades, J.C.; Grasa, G.; Alonso, M.; Rodriguez, N.; Aanthony, E.J.; Romeo, L.M. Cost structure of a postcombustion CO_2 capture system using CaO. *Environ. Sci. Technol.* **2007**, *41*, 5523–5527. [CrossRef] [PubMed]

70. Veltman, K.; Singh, B.; Hertwich, E.G. Human and Environmental Impact Assessment of Postcombustion CO_2 Capture Focusing on Emissions from Amine-Based Scrubbing Solvents to Air. *Environ. Sci. Technol.* **2010**, *44*, 1496–1502. [CrossRef] [PubMed]

71. Plaza, M.G.; Pevida, C.; Arias, B.; Fermoso, A.J.; Casal, M.D.; Martín, C.F.; Rubiera, F.; Pis, J.J. Development of low-cost biomass-based adsorbents for postcombustion CO_2 capture. *Fuel* **2009**, *88*, 2442–2447. [CrossRef]

72. Li, Y.; Su, M.; Xie, X.; Wu, S.; Liu, C. CO_2 capture performance of synthetic sorbent prepared from carbide slag and aluminum nitrate hydrate by combustion synthesis. *Appl. Energy* **2015**, *145*, 60–68. [CrossRef]

73. Li, Y.; Sun, R.; Liu, C.; Liu, H.; Lu, C. CO_2 capture by carbide slag from chlor-alkali plant in calcination/carbonation cycles. *Int. J. Greenh. Gas Control* **2012**, *9*, 117–123. [CrossRef]

74. Sun, J.; Liu, W.; Hu, Y.; Wu, J.; Li, M.; Yang, X.; Wang, W.; Xu, M. Enhanced performance of extruded–spheronized carbide slag pellets for high temperature CO_2 capture. *Chem. Eng. J.* **2016**, *285*, 293–303. [CrossRef]

75. Yu, J.; Le, Y.; Cheng, B. Fabrication and CO_2 adsorption performance of bimodal porous silica hollow spheres with amine-modified surfaces. *RSC Adv.* **2012**, *2*, 6784–6791. [CrossRef]

76. Qin, C.; Liu, W.; An, H.; Yin, J.; Feng, B. Fabrication of CaO-Based Sorbents for CO_2 Capture by a Mixing Method. *Environ. Sci. Technol.* **2012**, *46*, 1932–1939. [CrossRef] [PubMed]

77. Sun, L.B.; Li, A.G.; Liu, X.D.; Liu, X.Q.; Feng, D.; Lu, W.; Yuan, D.; Zhou, H.C. Facile fabrication of cost-effective porous polymer networks for highly selective CO_2 capture. *J. Mater. Chem. A* **2015**, *3*, 3252–3256. [CrossRef]

78. Liu, R.L.; Ji, W.J.; He, T.; Zhang, Z.L.; Zhang, J.; Dang, F.Q. Fabrication of nitrogen-doped hierarchically porous carbons through a hybrid dual-template route for CO_2, capture and haemoperfusion. *Carbon* **2014**, *76*, 84–95. [CrossRef]

79. Chandra, V.; Yu, S.U.; Kim, S.H.; Yoon, Y.S.; Kim, D.Y.; Kwon, A.H.; Meyyappan, M.; Kim, K.S. Highly selective CO_2 capture on N-doped carbon produced by chemical activation of polypyrrole functionalized graphene sheets. *Chem. Commun.* **2012**, *48*, 735–737. [CrossRef] [PubMed]

80. Garciagallastegui, A.; Iruretagoyena, D.; Gouvea, V.; Mokhtar, M.; Asiri, A.M.; Basahel, S.N.; AI-Thbaiti, S.A.; Alyoubi, A.O.; Chadwick, D.; Shaffer, M.S.P. Graphene Oxide as Support for Layered Double Hydroxides: Enhancing the CO_2 Adsorption Capacity. *Chem. Mater.* **2012**, *24*, 4531–4539. [CrossRef]

81. Li, X.; Cheng, Y.; Zhang, H.; Wang, S.; Jiang, Z.; Guo, R.; Wu, H. Efficient CO_2 Capture by Functionalized Graphene Oxide Nanosheets as Fillers to Fabricate Multi-Permselective Mixed Matrix Membranes. *ACS Appl. Mater. Interface* **2015**, *7*, 5528–5537. [CrossRef] [PubMed]

82. Shen, J.; Liu, G.; Huang, K.; Jin, W.; Lee, K.R.; Xu, N. Membranes with fast and selective gas-transport channels of laminar graphene oxide for efficient CO_2 capture. *Angew. Chem.* **2015**, *54*, 588–592. [CrossRef]

83. Li, Q.; Yang, J.; Feng, D.; Wu, Z.; Wu, Q.; Park, S.S.; Ha, C.S.; Zhao, D.Y. Facile synthesis of porous carbon nitride spheres with hierarchical three-dimensional mesostructures for CO_2, capture. *Nano Res.* **2010**, *3*, 632–642. [CrossRef]

84. Lu, J.; Zhang, J. Facile synthesis of azo-linked porous organic frameworks via reductive homocoupling for selective CO_2 capture. *J. Mater. Chem. A* **2014**, *2*, 13831–13834. [CrossRef]

85. Kim, H.; Jang, H.D.; Choi, M. Facile synthesis of macroporous Li_4SiO_4, with remarkably enhanced CO_2, adsorption kinetics. *Chem. Eng. J.* **2015**, *280*, 132–137. [CrossRef]

86. Zheng, B.; Xu, J. Carbon Capture and Storage Development Trends from a Techno-Paradigm Perspective. *Energies* **2014**, *7*, 5221–5250. [CrossRef]

87. Feldpauschparker, A.M.; Burnham, M.; Melnik, M.; Callaghan, M.L.; Selfa, T. News media analysis of carbon capture and storage and biomass: Perceptions and possibilities. *Energies* **2015**, *8*, 3058–3074. [CrossRef]

88. You, H.; Seo, Y.; Huh, C.; Daejun, C. Performance Analysis of Cold Energy Recovery from CO_2 Injection in Ship-Based Carbon Capture and Storage (CCS). *Energies* **2014**, *7*, 7266–7281. [CrossRef]

89. Hardisty, P.E.; Sivapalan, M.; Brooks, P. The Environmental and Economic Sustainability of Carbon Capture and Storage. *Int. J. Environ. Res. Public Health* **2011**, *8*, 1460–1477. [CrossRef] [PubMed]

90. Strazza, C.; Borghi, A.D.; Gallo, M. Development of Specific Rules for the Application of Life Cycle Assessment to Carbon Capture and Storage. *Energies* **2013**, *6*, 1250–1265. [CrossRef]

91. Lin, Q. The feasibilities and principles of climate negotiation guided by climate ethics. *Int. Forum* **2017**, *19*, 7–13. [CrossRef]

92. Qiu, H.-H.; Yang, J. An Assessment of Technological Innovation Capabilities of Carbon Capture and Storage Technology Based on Patent Analysis: A Comparative Study between China and the United States. *Sustainability* **2018**, *10*, 877. [CrossRef]

93. Alphen, K.V.; Noothout, P.M.; Hekkert, M.P.; Turkenburg, W.C. Evaluating the development of carbon capture and storage technologies in the United States. *Renew. Sustain. Energy Rev.* **2010**, *14*, 971–986. [CrossRef]

94. Zhang, X.; Fan, J.L.; Wei, Y.M. Technology roadmap study on carbon capture, utilization and storage in China. *Energy Policy* **2013**, *59*, 536–550. [CrossRef]

95. Sun, L.; Dou, H.; Li, Z.; Hu, Y.; Hao, X. Assessment of CO_2 storage potential and carbon capture, utilization and storage prospect in China. *J. Energy Inst.* **2017**. [CrossRef]

96. Shackley, S.; Waterman, H.; Godfroij, P.; Reiner, D.; Anderson, J.; Draxlbauer, K.; Flach, T. Stakeholder perceptions of CO_2, capture and storage in Europe: Results from a survey. *Energy Policy* **2007**, *35*, 5091–5108. [CrossRef]

97. Rodrigues, C.F.A.; Dinis, M.A.P.; Sousa, M.J.L.D. Review of European energy policies regarding the recent "carbon capture, utilization and storage" technologies scenario and the role of coal seams. *Environ. Earth Sci.* **2015**, *74*, 2553–2561. [CrossRef]

98. Dijkstra, J.W.; Walspurger, S.; Elzinga, G.D.; Pieterse, J.A.Z.; Boon, J.; Haije, W.G. Evaluation of post combustion CO_2 capture by a solid sorbent with process modeling using experimental CO_2 and H_2O adsorption characteristics. *Ind. Eng. Chem. Res.* **2018**, *57*, 1245–1261. [CrossRef]

99. Ge, B.S.; Wang, T.; Sun, H.X.; Gao, W.; Zhao, H.R. Preparation of mixed matrix membranes based on polyimide and aminated graphene oxide for CO_2 separation. *Polym. Adv. Technol.* **2018**, *29*, 1334–1343. [CrossRef]

100. Llerena, P.; Oltra, V. Diversity of innovative strategy as a source of technological performance. *Struct. Chang. Econ. Dyn.* **2002**, *13*, 179–201. [CrossRef]

101. Zhou, L.; Duan, M.; Yu, Y. Exergy and economic analyses of indirect coal-to-liquid technology coupling carbon capture and storage. *J. Clean. Prod.* **2018**, *174*, 87–95. [CrossRef]

102. Braun, C.; Merk, C.; Pönitzsch, G.; Rehdanz, K.; Schmidt, U. Public perception of climate engineering and carbon capture and storage in Germany: Survey evidence. *Clim. Policy* **2018**, *18*, 471–484. [CrossRef]

Article

Effect of the Implementation of Carbon Capture Systems on the Environmental, Energy and Economic Performance of the Brazilian Electricity Matrix

Claudia Cristina Sanchez Moore and Luiz Kulay *

Chemical Engineering Department, Polytechnic School, University of Sao Paulo, Avenida Professor Lineu Prestes, 580, Bloco 18—Conjunto das Químicas, São Paulo 05508-000, SP, Brazil; ccristina569@usp.br
* Correspondence: luiz.kulay@usp.br; Tel.: +55-11-3091-2233

Received: 14 December 2018; Accepted: 18 January 2019; Published: 21 January 2019

Abstract: This study examined the effect of Carbon Capture and Storage units on the environmental, energy and economic performance of the Brazilian electric grid. Four scenarios were established considering the coupling of Calcium Looping (CaL) processes to capture CO_2 emitted from thermoelectric using coal and natural gas: S1: the current condition of the Brazilian grid; S2 and S3: Brazilian grid with CaL applied individually to coal (TEC) and gas (TGN) operated thermoelectric; and S4: CaL is simultaneously coupled to both sources. Global warming potential (GWP) expressed the environmental dimension, Primary Energy Demand (PED) was the energy indicator and Levelised Cost of Energy described the economic range. Attributional Life Cycle Assessment for generation of 1.0 MWh was applied in the analysis. None of the scenarios accumulated the best indexes in all dimensions. Regarding GWP, S4 totals the positive effects of using CaL to reduce CO_2 from TEC and TGN, but the CH_4 emissions increased due to its energy requirements. As for PED, S1 and S2 are similar and presented higher performances than S3 and S4. The price of natural gas compromises the use of CaL in TGN. A combined verification of the three analysis dimensions, proved that S2 was the best option of the series due to the homogeneity of its indices. The installation of CaL in TECs and TGNs was effective to capture and store CO_2 emissions, but the costs of this system should be reduced and its energy efficiency still needs to be improved.

Keywords: electricity production; carbon capture; calcium looping; life cycle assessment; GHG mitigation

---

1. Introduction

According to information released by the International Panel on Climate Change (IPCC), global Greenhouse Gases emissions (GHG) in 2010 exceeded the 49 Gt CO_{2eq} mark. The most significant portion of this total originates from the generation of electric and thermal energy from fossil fuel burning. Prognoses by the same institution indicate that, by 2030, atmospheric CO_2 concentrations will reach between 600 and 1550 ppm, making the dynamic balance between anthropic systems and the biosphere unsustainable [1,2]. The Brazilian matrix differs slightly from this profile, given that 43% of the energy consumed in the country in 2017 originated from renewable sources. This behavior is strongly influenced by the domestic electricity supply, whose share of renewables was led by hydroelectric power plants, contributing with 65% of the generated total [3]. This model is, however, threatened in extreme situations, such as those recorded between 2012 and 2015, when effects as changes in rainfall regimes, associated with increases in demand (which was not fully supplied by hydropower), exposed Brazil to successive energy crises [4].

Phenomena such as population growth, urbanization, improved industrialization and increased availability of digital and computer systems over the past decade have increased the country's domestic electricity supply [3]. This scenario led the national electric system operator to seek short-term ways to enhance the capacity of the Brazilian electric grid (BR grid). This strategy has increased the participation of fossil sources, such as coal, petroleum derivatives and natural gas in the BR grid and, consequently, GHG emissions [5,6].

Faced with this situation, actions aimed at raising the electricity conversion efficiency and/or utilization rate, as well as the intensification of renewable source use, were established in order to improve the environmental performance levels of the system [7]. However, the discovery of gas and oil fields on the coastlines of the states of Rio de Janeiro and São Paulo, increasing energy generation potential from these resources, projects a scenario in which fossil sources will continue to represent a significant portion of the BR grid.

An opportunity to reconcile the domestic electricity supply expansion policy with the defined GHG emission reduction proposals is to adopt Carbon Capture and Storage (CCS) technologies. Such systems make it possible to separate the CO_2 generated in anthropic transformations that use combustion of fossils as way to generate energy to later store it in places where the CO_2 does not come in contact with the atmosphere [8–10].

Recent developments give CCS the status of a technically-economically viable alternative in terms of mitigating global warming. In this context, the study carried out by Cormos and Petrescu [11], who evaluated the suitability of the Calcium Looping (CaL) process as a carbon capture option in coal- and natural gas-operated electricity generation systems, is noteworthy. The authors observed significant decreases in atmospheric CO_2 emissions (66–82 kg/MWh) compared to plants where CCS were not implanted (760–930 kg/MWh). Following the same trend, Cormos [12] compared the performance of reactive absorption systems based on methyldiethanolamine and of CaL systems in capturing CO_2 in gas-powered plants. The author noted that the CCS approach provides better environmental and economic indicators than its counterpart for the study conditions.

Other researchers are more skeptical about the effectiveness of CCS in curbing global warming. In their view, these techniques are innocuous, as they only delay the release of CO_2 into the air, since this gas cannot be stored indefinitely [13–18]. Moreover, for these scientists, the processes typically applied by CCS—absorption, adsorption, and membrane separation—are also uneconomical. Thus, their implementation would raise the electricity unit price in situations where regulations or subsidies are not practiced. Adanez et al. [14] and Rochedo et al. [19] also demonstrated the intensive energy character of CCS technologies that can reduce the energy efficiency and operational flexibility of plants in which they are applied. There is, however, one wing of the scientific community that prefers to examine these arrangements more rigorously. Performing independent studies, Singh et al. [20], Korre et al. [21] and Branco et al. [22] noted that, even in cases where the use of CCS results in high CO_2 contents in combustion gases of natural gas- and coal-fueled thermoelectric plants, this effect will be attenuated due to GHG emissions occurring during other stages of the same process cycle. In order to reach this conclusion, the authors comprehensively determined the releases of GHG along the established productive arrangement, so that the CCS could fulfill the applications for which they were designed. This estimation was possible only through the Life Cycle Assessment (LCA) technique. Singh et al., Korre et al. and Branco et al. noted that by adopting the LCA approach—from which GHG emissions from energy and chemical consumption and infrastructure resources are also included in performance analysis—the efficiency of carbon sequestration systems could be drastically reduced, reaching limits of only 14% to 23%.

The literature records an expressive set of studies in which the application of CCS systems in fossil-fueled thermoelectric plants is addressed under different perspectives. However, a survey carried out with the same sources did not identify correlated research in which multiple analysis dimensions were verified by applying a systemic approach. This research seeks to fill this gap, even if only partially, by approaching the effects of the installation of CCS systems on the environmental, energy and economic performance of the Brazilian grid in a systemic way.

The findings of this initiative will, hopefully, serve as a basis for future Brazilian energy planning to address the trend antagonism established between the (expected) expansion of the domestic electricity supply and the (desirable) reduction of the global warming impacts that originate from electricity generation in Brazil.

2. Material and Methods

The method established for this study includes six steps: (i) characterization in terms of average technology, operational aspects, resource consumption and emissions from the current Brazilian grid; (ii) definition of electricity generation scenarios considering CCS implementation in coal and natural gas thermoelectric plants; (iii) estimate of the consumption of raw materials, inputs, process additives and utilities required by CCS systems integrated with thermoelectric plants; (iv) drawing up electricity generation Life Cycle Inventories coupled with CCS systems to quantify global warming impacts from GHG emissions and Primary Energy Demand throughout the life cycle of the defined arrangements; (v) accomplishment of an Economic Analysis to determine the costs associated with each scenario; and (vi) performance of a combined analysis of the dimensions assessed in the study, in order to verify synergies and discrepancies arising from this integration.

2.1. Backgrounds

2.1.1. Current Overview of the Brazilian Grid

The Brazilian electric generator complex is characterized by the grouping of large hydroelectric projects, which represent 96 GW of installed capacity [23–25]. The largest fraction of hydroelectric generation occurs in plants that use combined cycles, with an installed capacity of over 4.0 GW, and average yield of $\eta_{CC,H} \approx 55\%$. These units are concentrated in the Southeast region, which also presents the highest electric energy demand rate [26]. The Brazilian grid comprises a share of thermoelectric generation derived from natural gas, fuel oil and coal burning, representing 27 GW of the total available power. In most cases, coal-fired power plants operate under a subcritical cycle, and use Pulverized Coal Combustion (PCC) systems, $\eta_{PCC} \leq 40\%$ [23]. Finally, the electric generation from petroleum derivatives is carried out in boilers or internal combustion engines. The fuels regularly used in these situations are fuel oil and diesel [23].

Another 14 GW of the Brazilian grid's installed comes from biomass-powered thermoelectric plants. Typically, the technological routes of this generation source comprise steam generator cycles with backpressure and condensation-extraction turbines, as well as a combined cycle integrated with biomass gasification. Although different types of fuel can be used in the process (i.e., wood, crop residues from rice, soy, and corn, or even urban and industrial waste), about 80% of the biomass consumed in Brazil for energy purposes is derived from bagasse and straw of sugarcane [27]. Boiler-turbine assemblies that operate according to the Rankine or Brayton cycles and produce steam at 65 bar and 550 °C have been extensively applied in electricity exports [23,26–28].

Brazilian nuclear generation comes from two plants: Angra I, comprising 640 MW of installed capacity, and Angra II, with capacity for 1350 MW. Located in the state of Rio de Janeiro, these plants are of the Pressurized Water Reactor (PWR) type [29]. Brazil has 534 wind farms distributed mostly in the states of Rio Grande do Norte, Bahia, Ceará, Rio Grande do Sul and Piauí. This generation source has significant expanse since 2013, reaching 13 GW of installed capacity in the current grid. Wind energy is seen as a promising option for future DESs, as it displays lower environmental impacts than some of its congeners [24,25].

Finally, by mid-2018, the country registered over 30,000 photovoltaic solar generation facilities, which together make up 2.4 GW of capacity [30]. Even though power indices are less significant than those from other renewable sources, the number of solar micro-generators grew by 407% from 2015 to 2017. This behavior can be attributed to technology maturing, lower investment costs, environment awareness by the population and fiscal and political incentives provided by the Brazilian government [23,29]. Figure 1 depicts the Brazilian installed capacity of electricity generation in the year of 2017 [30].

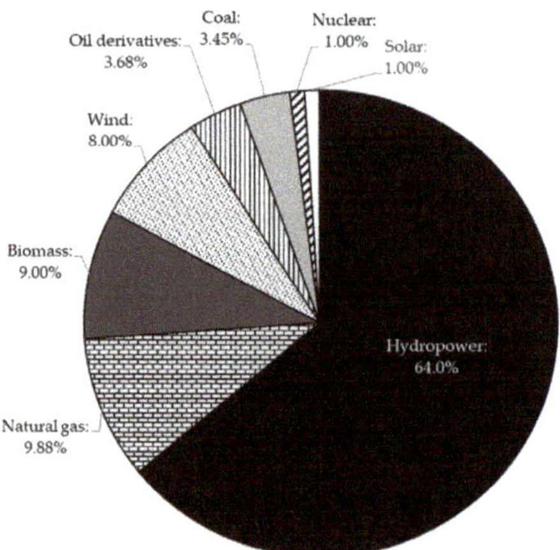

Figure 1. Brazilian installed capacity of electricity generation in 2017: relative distribution by source.

2.1.1.1. Carbon Capture and Storage and Calcium Looping Technology

Given their CO_2 generation volumes, CCS systems have spread more significantly in the energy, petrochemical, cement and metallurgical sectors [8,9,31]. In the energy sector such technologies are even seen as a means to enable the continuous use of fossil resources—thus avoiding their substitution by renewable sources—despite GHG emission levels [32,33].

In this study, CCS systems were represented by the CaL process. This decision was based on the following factors: (i) technological concept compatibility between the CaL process and the active thermoelectric plants in the country, which allows for system implementation without requiring significant adjustments to the electric generation power plant; (ii) reduced energy consumption; and, (iii) lower implementation and operating costs than those estimated for equivalent technologies [34,35].

In the CaL process, the CO_2 present in the combustion gases emanating from the plant reacts with calcium oxide (CaO) inside a vessel (carbonator) to form $CaCO_3$. This transformation is represented by Equation (1). The fact that it is endothermic predisposes the transformation to occur at temperatures ranging from 500 to 650 °C. The product stream from this stage feeds another reactor (calciner) in which CaO regeneration occurs, again by a thermal effect (Equation (2)). The conditions required for decomposition (800–950 °C) are provided by the burning of fossil fuels in the presence of pure O_2. Equation (3) represents the global combustion reaction (on stoichiometric bases) that occurs in a coal-fired power plant. Equation (4) describes the same transformation for a thermoelectric unit operated with natural gas [11,35,36]:

$$CaO + CO_2 \rightarrow CaCO_3 \tag{1}$$

$$CaCO_3 \rightarrow CO_2 + CaO \tag{2}$$

$$C + O_2 \rightarrow CO_2 + Heat \tag{3}$$

$$CH_4 + 2O_2 \rightarrow CO_2 + 2H_2O + Heat \tag{4}$$

The CO_2 is then dried and compressed prior to storage. The CaO, however, returns to the first reactor in order to be reused by the process [37].

2.2. Scenario Definition

Table 1 describes the scenarios selected for the analysis. The adopted system for this process was based on the application of four criteria. The first determined the creation of a reference scenario (S1) representing the current Brazilian grid arrangement. The second criterion established that the CaL process would be coupled only to coal and natural gas-operated thermoelectric plants. This decision was made taking into account the (high) Domestic Electricity Supply expansion potential of these sources [29].

Table 1. Specification of the analysis scenarios by type of electricity generation source

Source	Domestic Electricity Supply by Source (%)	Scenario			
		S1 (Baseline)	S2	S3	S4
Hydropower	65.2	C	C	C	C
Oil	2.51	C	C	C	C
Coal	4.09	C	CaL	C	CaL
Natural Gas	10.5	C	C	CaL	CaL
Biomass	8.23	C	C	C	C
Nuclear	2.52	C	C	C	C
Wind	6.82	C	C	C	C
Solar	0.13	C	C	C	C

Legend: (C): Technology currently practiced for the respective generation source; (CaL): indicates coupling with the CaL process.

The third criterion established that the environmental, energy and economic effects of CaL coupling would be verified both individually—by generation source (S2 and S3) –, and associated, in which case this CCS technology would be applied simultaneously to gas- and coal-powered plants (S4). The last criterion defined that the other generation sources constituting the Brazilian grid would be represented by technologies consistent with those currently practiced in the country (Section 2.1.1). This measure sought to avoid the influence of external parameters to the field of analysis on the research findings. The domestic electricity supply data per electricity generation source were obtained from the National Energy Balance for 2018, for a total production of 588 TWh [38].

2.3. Calcium Looping Process Simulation: Assumptions and Process Parameters

The CaL behavior was simulated with the aid of the Aspen Plus software v.8.8 by Aspentech®. The process, like others involved in the simulation, operates in steady state. Production patterns and technical parameters adopted for these estimates agree with the average technological profile practiced in Brazil for the corresponding modeled plants. In the case of coal-fired thermoelectric plants, the CaL was coupled to a plant with an electric generation capacity of 500 MW, equipped with a system for the removal of sulfur compounds from the combustion gases (flue gas desulfurization), whose average efficiency reaches $\eta_S = 90\%$. It was assumed by boundary condition that the combustion gas would be fed to the carbonator only after it had been cooled and desulphurized. Bituminous coal was considered as 'non-conventional' flow, due to its composition profile, represented in Table 2 [34].

Table 2. Analyses of the bituminous coal used in the simulation.

Ultanal (%$_{w/w}$)	
C	72.3
H	4.11
O	5.93
S	0.58
N	1.70
Cl	0.00
Ash	15.4
LHV (MJ/kg)	22.2

Table 2. *Cont.*

Proxanal (%$_{w/w}$)	
moisture	8.00
volatiles	24.9
fixed carbon	59.7
ash	15.4
Sulfanal (%$_{w/w}$)	
sulfate	0.26
pyritic	0.26
organic	0.06

Legend: Ultanal: Ultimate analysis; Proxanal: Proximate analysis; Sulfanal: Sulfur sources analysis; LHV: Low Heat Value.

For the thermoelectric plants that run on natural gas, the defined installed capacity was of 300 MW. The molar average composition of the fuel used in the simulation is given in Table 3 below [39].

Table 3. Average composition of natural gas.

Component	CH$_4$	C$_2$H$_6$	C$_3$H$_8$	C$_4$H$_{10}$	CO$_2$	N$_2$	Others
Molar fraction (%)	86.1	8.15	2.14	0.52	0.72	1.34	1.03

The process diagrams, corresponding to the coal-fired and natural gas thermoelectric power plants are presented in Figure 2a,b. The carbonator and calciner behaviors were simulated by RStoic-type reactors, a model available in Aspen Plus® software to represent transformations whose progression profile is stoichiometric or close enough to it. The rate of CO_2 conversion into $CaCO_3$ in the carbonator was determined by Equation (5):

$$E_{CO2} = \left(\frac{F_R}{F_{CO2}} \right) \times X_{carb} \tag{5}$$

where (X_{carb}) corresponds to the conversion rate of CO_2 to $CaCO_3$ in the carbonator, (E_{CO2}) refers to the CO_2 removal efficiency of the combustion gas fed to the carbonator. (F_R / F_{CO2}) refers to the molar ratio between the CaO that circulates between this device and the calciner (F_R) and the CO_2 current introduced into the carbonator (F_{CO2}). The conversion rate (η_r) in the calciner, which expresses the degree of CaO regeneration, was set at $\eta_r = 100\%$.

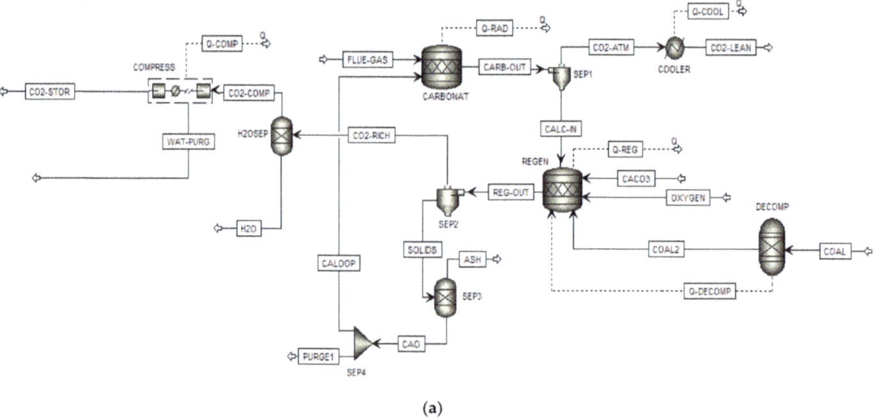

(a)

Figure 2. *Cont.*

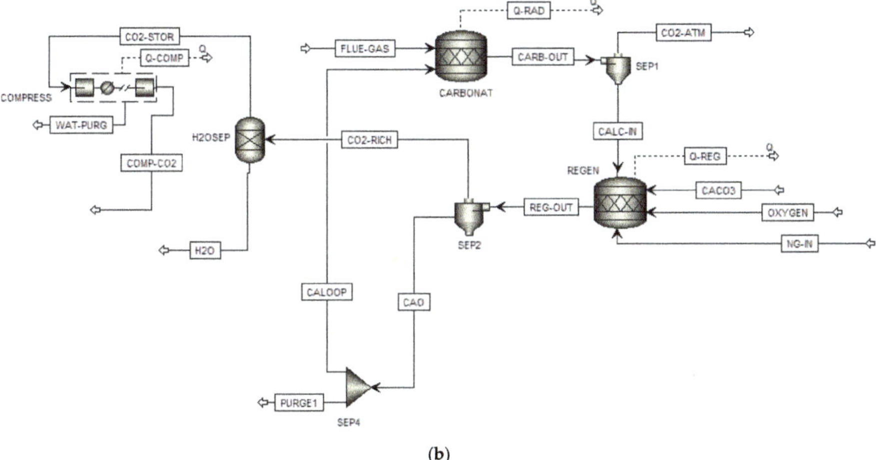

(b)

Figure 2. Calcium Looping system (**a**) coupled to a coal-fired thermoelectric plant, and; (**b**) coupled to a natural gas -fired thermoelectric power unit.

The oxygen burned for the heat generation inside the calciner is obtained from the Air Separation Unit (ASU). Oxygen flows and the fuel consumed in the calciner (coal or natural gas) have been determined using the Design Spec tool available in the Aspen Plus® software. Pure CO_2 is raised to a pressure of 80 bar, after being subjected to three compression stages interspersed by chillers. The gas is cooled to 28 °C during each of these stages [40]. The basic process parameters used for the simulation of the CaL system coupled to the coal-fired and natural gas power plants are presented in Table 4 [33,40–42].

Table 4. Process parameters that specify the Calcium Looping system for each type of thermoelectric plant.

Parameter	Coal-Fired Thermoelectric Plant	Natural Gas Thermoelectric Plant
Fuel used in the calciner	Coal	Natural Gas
Carbonator Temperature (°C)	650	600
Carbonator Pressure (bar)	1.00	1.00
Calciner Temperature (°C)	900	900
Calciner Pressure (bar)	1.00	1.00
E_{CO2} (%)	90.0	90.0
(F_R/F_{CO2})	7.00	14.0
(F_{CaCO3}/F_{CO2})	0.10	0.10
CaO Purge	0.10	0.10
Fresh $CaCO_3$ temperature (°C)	25.0	25.0
O_2 purity degree (%v/v)	95.0	95.0
O_2 stream temperature (°C)	15.0	15.0
Fuel temperature (°C)	25.0	25.0

2.4. Environmental Assessment

Environmental performance was determined according to the Life Cycle Assessment approach. The LCA is a management approach capable of quantifying the environmental impacts provided by a product (or service) throughout its life cycle; i.e., within a domain that covers raw material extraction and processing operations, the manufacturing chain of the product, its use, recycling and final disposal, as well as the transport and distribution operations to which it is submitted to [43].

The LCA methodology comprises four structural phases: (i) Goal and Scope definition; (ii) Life Cycle Inventory; (iii) Impact Assessment, and (iv) Result Interpretation [44]. This study followed the conceptual guidelines and requirements described in the ISO 14044 standard [45], with the LCA carried out according to the attributional approach and a "cradle-to-gate" application scope. The environmental impacts that originate in each scenario were measured by a Reference Flow (RF) of: '*to generate 1.0 MWh of electricity*', from the sources and proportions that compose the Brazilian grid for 2018 (Table 1). Figure 3 presents the elements of each model (Product System) created to describe the study situations, in a generic form.

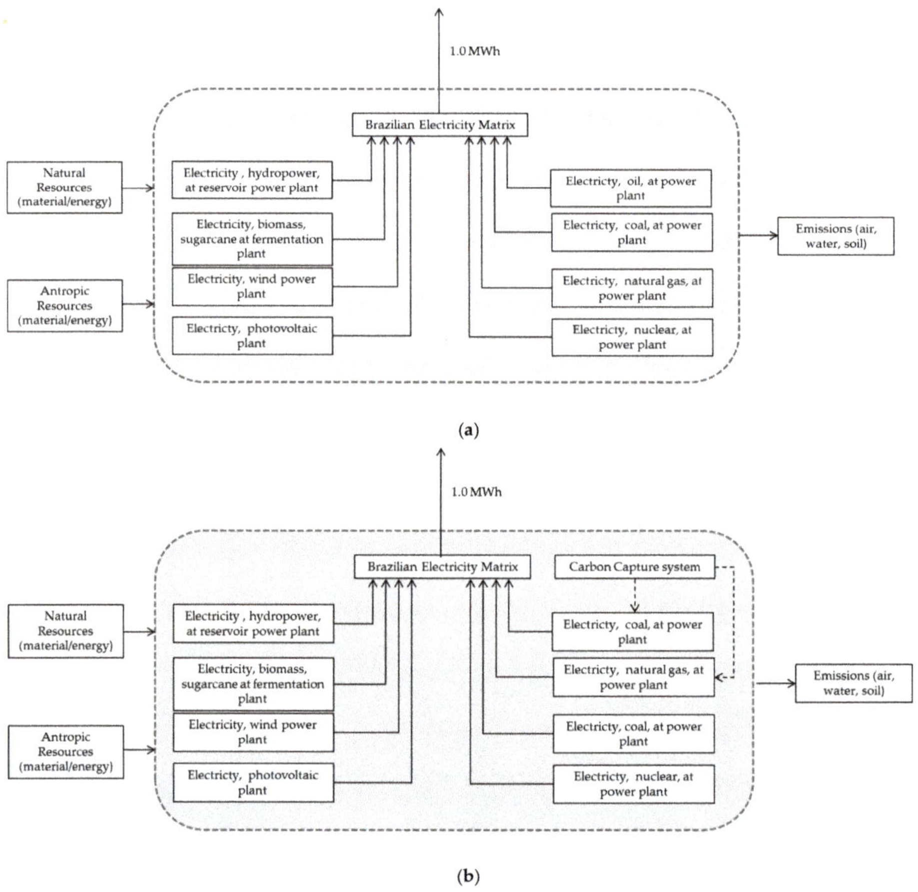

Figure 3. Product System for the production of 1.0 MWh of electricity: (**a**) BR grid in its original structure, and; (**b**) coupled to CCS.

Life Cycle Inventories (LCIs) were drawn up with the support of SimaPro - Pre-Consultants® computer software from secondary data obtained from the Ecoinvent database [46]. This database provides LCIs of average electricity generation technologies whose characteristics are representative the Brazilian electricity generation systems, and in some cases were adapted to Brazilian conditions. For hydroelectric generation, the conversion efficiency considered by the study was of η_{HY} = 95%, while for thermoelectric plants operating with coal and fuel oil, performances were of η_{Coal} = 40% and η_{FO} = 37%, respectively. In the case of thermonuclear and thermoelectric plants using natural gas, efficiencies were set at η_{NU} = 33% and η_{NG} = 38% respectively. The consolidated LCI for Brazilian electricity production were elaborated respectively from the following databases: (i) Hydropower:

"*Electricity, hydropower, at reservoir power plant/BR U*"; (ii) Oil: "*Electricity, oil, at power plant/CH U*", considering an average net efficiency of $\eta_{FO} = 36\%$; (iii) Coal: "*Electricity, hard coal, power plant/UCTE U*", with an average net efficiency of $\eta_{PCC} = 40\%$; (iv) Natural Gas: "*Electricity, natural gas, at power plant/UCTE U*" adapted to the Brazilian natural gas supply conditions and the respective logistics of resource distribution, for an average net efficiency of $\eta_{NG} = 38\%$; (v) Nuclear: "*Electricity, nuclear, at power plant/CH U*" admitting only the generation in systems with reactors type PWR; (vi) Solar: "*Electricity, photovoltaic production mix, at plant/CH U*" and (vii) Wind: "*Electricity at wind power plant/RER U*". All these assemblies have been sufficiently modified to depict the local operating conditions of these sources. For Biomass, the LCI "*Electricity, bagasse, sugarcane, at fermentation plant/BR U*" used in this study was adapted from premises established by Guerra et al. [28].

For the scenarios that consider the CaLcoupled to coal- and natural gas (i.e., S2—S4) thermoelectric plants, the LCIs from the Ecoinvent database were adjusted and supplemented with data on the consumption of raw materials and inputs and emissions, which occur in those processes, obtained from the Aspen Plus® simulations (Section 2.3). Interactions with the $CaCO_3$ processing surroundings were represented by a version adapted to the national conditions of the LCI "*Limestone, milled, loose, at plant/CH U*". The electrical consumption of the ASU plant (200 kWh/t O_2) and the compressor arrangement on purified CO_2 (100 kWh/t CO_2) were also assess [41,47].

Concerning the preparation transport LCIs for $CaCO_3$, it was assumed that: (i) displacements occur by road in trucks with cargo capacity between 16–32 t, which represent Brazilian conditions [48]; (ii) the average distance between the extraction mine (located in Northeastern Brazil), and the coal-operated thermoelectric plant (South Brazil) was of 4408 km. For the natural gas thermoelectric plant (installed in the Southeast), displacement was of 3000 km.

The ReCiPe midpoint (H) v 1.12 [49] method was applied to quantify impact potentials in terms of global warming. In general terms, this quantification occurs multiplying the totalized amount of each GHG by its respective impact factor (IF). The Impact factors—coefficients that describe the magnitude of GHG effects in terms of global warming—adopted by ReCiPe midpoint (H) originate from scientific research and development conducted by the Intergovernmental Panel on Climate Change (IPCC). According to this approach, CO_2 was defined as the reference substance for category, thus receiving IF = 1.0 kg $CO_{2\,eq}$/kg CO_2 emitted. The coefficients for CH_4 and N_2O are estimated based on this standard, and for ReCiPe v1.12, they correspond respectively to 25 kg $CO_{2\,eq}$/kg CH_4 and 298 kg $CO_{2\,eq}$/kg N_2O.

The energy impacts of each scenario were estimated in terms of Primary Energy Demand by the Cumulative Energy Demand (CED) v1.10 method. CED describes the impacts related to depletion of the Earth's primary energy for both non-renewable (fossil, nuclear, biomass) and renewable (biomass, wind, solar, and geothermal and water) energy resources [50]. For this, the method applies a conceptual logic similar to that used by ReCiPe for global warming, of multiplication of impact precursors by their respective IFs. In CED, the intrinsic energy content of a natural resource is used to estimate the primary energy demand associated with it, throughout the life cycle of a good (or service). This intrinsic energy refers, therefore, to the impact factor of the resource. For fossil fuels and biomass, IF is represented by the Higher Heating Value. Regarding nuclear energy this index is based on the uranium chain, considering the characteristics of the average German pressurized water reactor. For hydropower an IF = 1.00 MJ/MJ is assumed, despite the energy precursor of impact (potential, from hydropower or even from hydrogen). On the other hand, if Primary Energy Demand comes from water stored in a barrage pond, IF = 10.0 kJ/kg water. Finally, for other renewable sources (wind, solar, geothermal), CED admits that energy input equals the amount of energy converted [50].

2.5. Economic Analysis

The economic analysis is based on the Levelised Cost of Energy (LCOE) method [51]. Such an approach consists of one of the ways to instrumentalize the variables involved in the economic dimension by making different sources of electricity generation comparable. The LCOE corresponds to the specific real cost (per kilowatt-hour) relative to the construction and operation of the plant, within the time horizon—the year (*t*)—for which it was designed, including maintenance actions. This indicator can also be defined as the specific average revenue required, measured per unit of produced energy, so that the entrepreneur can recover operation and maintenance investments and expenses that affect the project [51]. The expression that determines the LCOE is represented in Equation (6):

$$LCOE = \frac{\sum(Capital_t + O\&M_t + Fuel_t + Carbon_t + D_t) \times (1+r)^{-t}}{\sum MWh\,(1+r)^{-t}} \tag{6}$$

($Capital_t$) indicates the total construction costs in year (*t*), the ($O\&M_t$) factor refers to operation and maintenance expenditures, ($Fuel_t$) is the fuel cost and ($Carbon_t$), the carbon cost. (D_t) rate depicts decommissioning and waste management costs and (($1+r$)$^{-t}$) represents the discount factor.

The components used to determine LCOE include capital, fuel, operation and maintenance, and financing costs. In addition, technical aspects of the processes under analysis (i.e., plant efficiency and capacity factor) are also considered by the estimate [51].

For solar and wind technologies that dispense fuel consumption, and present discrete operating and maintenance expenses, the LCOE value is largely conditioned to the estimated cost of capital for generation capacity.

When fuel costs are high, as occurs in oil- and natural gas-operated thermoelectric plants, the LCOE value tends to vary significantly. As with any projection, uncertainties associated with the components of the estimation are noted, which are generally introduced due to geographic aspects, degree of electricity generation technology consolidation and fuel price quotations [52].

For this study, the LCOE values for each scenario ($LCOE_i$) were determined on the basis of data and information contained in International Energy Agency (IEA) and EPE documents for an annual discount rate r = 10%, used for high-risk market investments, which was considered constant throughout the life cycle of the enterprise [52,53]. The input parameters for the LCOE calculation are presented in Table 5.

Table 5. Parameters for the LCOE calculation for each energy source for a discount rate of r = 10% (USD/MWh).

Source	Capital Cost	O&M Cost	Fuel Cost	Carbon Cost	Decommissioning Costs
Hydropower	50.9	9.00	0.00	0.00	0.04
Oil	40.5	7.56	149	23.3	0.03
Coal	40.5	7.56	30.8	23.3	0.03
Natural Gas	15.9	5.72	74.6	10.3	0.03
Biomass	86.6	17.6	93.2	0.00	0.08
Nuclear	74.4	13.1	10.0	0.00	0.12
Wind	94.0	23.0	0.00	0.00	0.62
Solar	140	27.9	0.00	0.00	0.00

For Solar energy, O&M costs include decommissioning costs. For oil-fired thermoelectric plants, it was decided that Capital, O&M, Carbon and Decommissioning costs would be the same as those of the gas-fired thermoelectric plant, assuming an oil cost twice that of natural gas.

2.6. Combining Environmental, Energy and Economic Indicators

The research also sought to determine the combined effect of the analyzed dimensions on the electric generation trends from the different energy sources that constitute the Brazilian grid, with and without CCS system coupling. Thus, environmental, energy and economic performance indices were calculated by normalizing the maximum value of the results obtained in previous stages of the study. This procedure is described below by Equations (7)–(9):

$$EI_i = \left(\frac{I_{GWP,i}}{I_{GWP,S1}} \right) \tag{7}$$

$$EnI_i = \left(\frac{I_{PED,i}}{I_{PED,S1}} \right) \tag{8}$$

$$EcI_i = \left(\frac{I_{LCOE,i}}{I_{LCOE,S1}} \right) \tag{9}$$

where: (EI_i): environmental indicator for scenario i; ($I_{GWP,i}$): impact in terms of Global Warming Potential for scenario i; ($I_{GWP,\,S1}$): the impact result for S1 in terms of Global Warming Potential; (EnI_i): energy indicator for scenario i; ($I_{PED,i}$): impact in terms of Primary Energy Demand for scenario i; ($I_{PED,S1}$): the impact result for S1 in terms of Primary energy Demand among all scenarios; (EcI_i): economic indicator for scenario i; ($I_{LCOE,i}$): the Levelised Cost of Energy for scenario i; ($I_{LCOE,S1}$): the value of Levelised Cost of Energy corresponding to S1.

A Combined Indicator (CI_i) was calculated for each scenario relating the normalized indicators, as described in Equation (10):

$$CI_i = (EI_i \times EnI_i \times EcI_i) \tag{10}$$

3. Results and Discussion

According to the obtained results, in order to achieve a 90% capture efficiency of the CO_2 present in the exhaust stream of a coal-fired thermoelectric plant coupled to a CaL-type CCS system, a 48.6 kg/s coal stream to supply energy to the calciner is required. In the case of the natural gas thermoelectric plant, the fuel addition rate is of 23.0 m^3/s. In addition, $CaCO_3$ consumption in the coal-fired thermoelectric was of 34.9 kg/s while the expenditure in the natural gas technology was of only 10.0 kg/s.

The thermal energy recovered from the carbonator due to the exothermic reaction and the gas streams leaving the compressor arrangement and calciner totaled 1370 MW for the coal-fired thermoelectric units and 600 MW for the natural gas power plants. The use of this energy allowed for an increase in gross power in both thermoelectric plants to, respectively, 3000 MW and 1400 MW. Of these totals, however, 175 MW and 62 MW should still be discounted, as they correspond to the CaL demands for each plant. The net efficiencies obtained for coal-fired thermoelectric and natural gas power plants were, respectively, of η_{Coal} = 38% and η_{NG} = 34%.

These values are considered acceptable due to their low variability when compared to equivalent plant indices without CCS of, respectively, 40% and 38% [46]. These performances demonstrate the technical advantages of implementing the CaL process, despite the thermoelectric plant operates with coal or natural gas. The results with this technology were even higher than those obtained by other CCS systems in similar situations, such as that using monoethanolamine solvents, for $\eta_{Coal} \approx \eta_{NG} \approx$ 28% [22,54].

3.1. Environmental and Energy Analyses

Table 6 depicts the cumulative impact values for Global Warming Potential and Primary Energy Demand, as well as their main precursors for each analysis scenario. These results indicated that there was no scenario that accumulated simultaneously, the best performance of the whole series in both dimensions of analysis.

In terms of global warming, it is noted that CO_2 and CH_4 are the precursors displaying the major impact, and that emissions from other sources are only slightly changed as a function of the CaL coupling to thermoelectric plants. This is explained by the fact that $CH_{4,b}$, $CO_{2,\,LT}$ and N_2O are associated with hydroelectric sources and a biomass-driven thermoelectric plant whose operation was not affected by the presence (or not) of the carbon capture system. These tendencies can also be observed in Figure 4 depicts how the impacts for Global Warming Potential are distributed by source of the Brazilian electric grid in both relative and absolute terms.

Table 6. Environmental and Energy Performance and main specific contribution for the of 1.0 MWh of electricity using the Brazilian grid.

Impact Category	Precursor	Scenario			
		S1	**S2**	**S3**	**S4**
Global Warming Potential (kg CO_{2eq})	CO_2	123	92.7	76.8	46.7
	$CO_{2,\,LT}$	73.2	73.2	73.2	73.3
	CH_4	55.1	55.2	81.0	81.1
	$CH_{4,b}$	28.7	28.7	28.8	28.8
	N_2O	3.16	3.02	3.13	3.01
	Total	283	253	263	233
Primary Energy Demand (MJ)	Crude Oil	344	392	406	454
	Natural Gas	1280	1290	1930	1940
	Coal	464	465	467	469
	Kinetic energy (water)	2484	2485	2488	2489
	Total	4572	4632	5291	5352

Legend: $CO_{2,\,LT}$: Carbon dioxide from land transformation; $CH_{4,b}$: Methane, biogenic.

The fact that S4 achieved the best global warming result of the whole series does not come as a surprise, considering that S2 and S3 had superior performances compared to S1. This is because S4 accumulates the effects, which, for GWP, are positive, of the implementation of the CaL process in coal-fired and natural gas power plants. However, S4 indicates a 17% reduction in the total GWP impact compared to S1. This is due to the 62% decrease in total CO_2 emissions from coal and gas burning that are used in these systems. If this stage of the process contributed to about 32% of S1 impacts, the presence of CaL reduces the relative share of combustion in the coal-fired and natural gas power plants to 3.0% of the total GWP impact of S4. Conversely, impacts from CH_4 emissions occurring during natural gas extraction and processing increased by 47% in the passage from S1 to S4. This is due to the increased consumption of the same fuel to meet CaL energy needs.

For S2, the incorporation of the CaL process in the coal-fired power plants avoided the emission of 30.0 kg CO_2/RF, generating an 11% reduction in GWP impacts in relation to the baseline scenario. Most CO_2 emissions result from natural gas burning at natural gas power plants, a stage contributing with 22% of the total impact of S2. On the other hand, CH_4 losses remained unchanged in the S1 → S2 passage.

In S3, the implementation of the CaL to mitigate CO_2 emissions from natural gas combustion reduced GWP impacts by 7.1%. In this context, gas extraction and processing had a significant influence on the cumulative impact, contributing with 33%. Coal burning at power plants accounted for 13% of the cumulative impact on the scenario. Finally, as discussed previously, the increase in CH_4 losses significantly dampened the advantage obtained by the decreased CO_2 losses accumulated by S3 in comparison to S1.

In addition to the adverse effects mentioned above, the advantages of the simultaneous installation of CaL processes for carbon capture in coal-fired power plants and natural gas power plants were also neutralized by inherent aspects of the technology itself. This is the case of $CaCO_3$ transport being fed into the regenerator for the purpose of make-up. This operation contributed, in isolation, to 2.57 kg CO_{2eq}/MWh for S2 and 3.03 kg CO_{2eq}/MWh for S3, reaching a 12% of the total contributed by CO_2 in S4, when the effect of the use of CaL was verified in both fronts. The impossibility of shared transport with other assets, the prevalence of the road mode for the distribution of heavy loads in Brazil, and, especially, the long distances between the $CaCO_3$ extraction mines and the plants (from 660 to 1100 km) justify these performances.

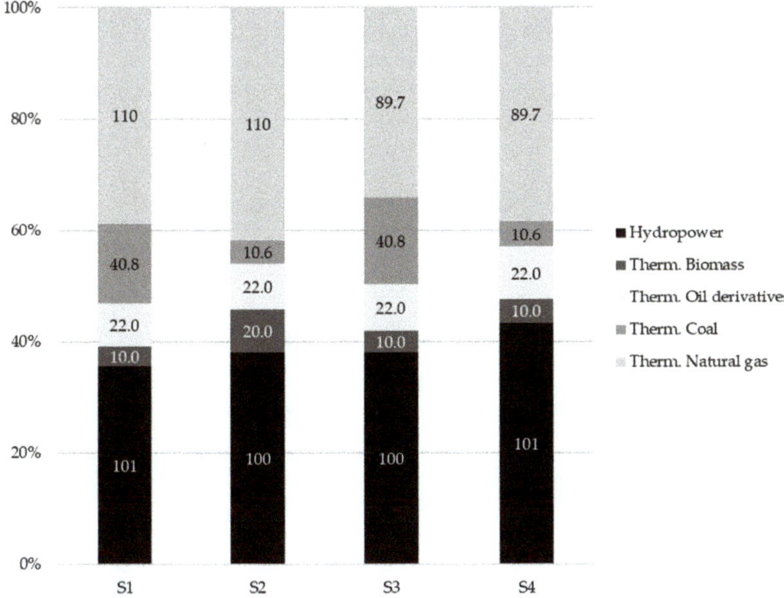

Figure 4. Contributions for Global Warming Potential by source of the Brazilian grid: relative (%) and absolute (kg CO_{2eq}/MWh) values.

The results obtained for Primary Energy Demand suggest a slight advantage of S1 over S2. In the baseline scenario, hydroelectricity production accounts for 54% of the total impact, followed by primary energy consumption in the form of natural gas, at 28% contribution. S2 closely follows these same trends, and the differences in contribution between the natural gas, coal and kinetic energy scenarios are tenuous. On the other hand, a 14% increase in crude oil consumption in the s1 $\rightarrow$ s2 passage as a consequence of the additional diesel consumption of to transport $CaCO_3$ was noted, generating a 1.3% increase in the total impact to S2 compared to S1.

S3 and S4 showed quite divergent profiles in relation to S1. These findings reflect negatively on the use of CaL for the natural gas power plants. A 16% increase in S3 impacts due to additional crude oil (18%) for diesel production used in $CaCO_3$ transport, and natural gas (51%) demands, was observed. For S4, the increase in total impact was of 17%, resulting in the worst energy performance among the evaluated options. This was, once again, due to increases in crude oil and natural gas consumption, which, in this situation, were of 32% and 52%.

3.2. Economic Analysis

Table 7 presents the Levelised Cost of Energy (LCOE) values discretized per constituent source of the BR grid and for the circumstances described in S1. The results indicate fuel oil burning as the main source of unit costs of electric generation in Brazil. The counterpart of this effect by hydroelectricity—a major participation source in the grid – allows to explain Brazil's position in the group of countries with the lowest aggregate electricity costs among industrialized countries [52].

Table 7. LCOE per source for a discount rate of 10% (USD/MWh).

Energy Source	Hydro	Oil	Coal	Natural	Biomass	Nuclear	Wind	Solar
LCOE (USD/MWh)	59.9	221	95.7	107	154	98.0	112	168

The LCOE value for the cases in which the CaL process is coupled to the coal-fired and natural gas power plants was based on estimates made by Mantripragada et al. [54]. The authors determined that the implantation of such a system in coal-fired power plants causes increases the LCOE in 137% in relation to a plant that does not present the same conditions. This percentage increase was applied to the LCOE values presented in Table 6 for the coal-fired and natural power plants. Table 8 presents an overview of electricity generation and costs (individualized by source) for each scenario.

Table 8. Electricity generation and economic performance by source and scenario.

Energy Source	Generation MWh/y	LCOE by Source (USD/MWh)	Electric Generation Cost (Billion USD/y)			
			S1	S2	S3	S4
Hydro	383×10^6	59.90	22.9	22.9	22.9	22.9
Oil	14.7×10^6	221	3.25	3.25	3.25	3.25
Coal	24.1×10^6	95.7	2.31	–	2.31	–
Coal + CaL	24.1×10^6	227	–	5.46	–	5.46
Natural Gas	61.7×10^6	107	6.60	6.60	–	–
Natural Gas + CaL	61.7×10^6	254	–	–	15.7	15.7
Biomass	48.2×10^6	154	7.42	7.42	7.42	7.42
Nuclear	14.7×10^6	98.0	1.44	1.44	1.44	1.44
Wind	40.0×10^6	112	4.48	4.48	4.48	4.48
Solar	7.64×10^6	168	0.13	0.13	0.13	0.13
Total	588×10^6	–	48.5	51.7	57.6	60.8

It is noted that the use of a CaL-type carbon capture system in natural gas thermoelectric plants (S3) results in higher costs than if the same system were to be coupled to a coal-fired power plant (S2). This is mainly due to the price of natural gas (74.60 USD/MWh), which is about two-fold of that associated with coal [52]. The magnitude of the coal price in the international market also justifies the smoothing of the cost increase of S2 (6.6%) in relation to S1. For the reasons given above, the cost variations of S1 → S3 and S1 → S4 were of 19% and 25%, respectively.

3.3. Combined Analysis: Environmental, Energy and Economic

The fact that no prevalent scenario for all dimensions was observed emphasizes the importance of a combined analysis. Table 9 presents the results of this action. According to this metric, S2 stands out from the others when achieving the best combined indicator (CI_{S2}). This finding indicates that the use of Calcium Looping (CaL) process in coal-operated thermoelectric plants only shows a feasible possibility by associating all the analysis dimensions.

Even with performance above S4 in environmental terms (7.8%), and of energy and economic order of S1 (1.3% and 6.5%), S2 dominates over its competitors due to the homogeneity of their indices. Conversely, S3 presented the worst combined performance, without even accumulating a lower individual result. Therefore, the application of CaL for CO_2 removal from the natural gas-fired thermoelectric plant emissions is inadvisable.

Table 9. Normalized Indicators and combined results for scenarios S1–S4.

Indicator	Scenario			
	S1	S2	S3	S4
EI_i	1.000	0.894	0.929	0.823
EnI_i	1.000	1.011	1.139	1.151
EcI_i	1.000	1.066	1.188	1.254
CI_i	1.000	0.963	1.257	1.188

The overall result achieved by S4 should also be evidenced. In this case, the setbacks to the energy and economic dimensions, when EnI_{S4} and EcI_{S4} occupy the last positions in their respective series, were counteracted by EI_{S4} performance. Thus, the simultaneous installation of CaL in coal-fired and natural gas operated power plants is a costly and energy-intensive alternative. Conversely, the effectiveness with which this technology fulfills its purposes of capturing and storing carbon emissions—in particular CO_2—is sufficiently elucidated to return it to the set of possibilities to be considered in a process decision-making process within the scope of Brazilian energy planning.

4. Conclusions

This study evaluated the effect of the implementation of Carbon Capture and Storage systems on the environmental, energy and economic performance of the Brazilian electrical matrix. Four scenarios were established considering the coupling of CaL processes to capture CO_2 emitted from thermoelectric plants using coal and natural gas. S1 describes the current condition of the Brazilian grid and is, therefore, defined as the baseline scenario. S2 and S3 refer to the BR grid with CaL applied individually to coal- and gas- operated thermoelectric plants, while in S4 the CCS technology is simultaneously applied to coal-fired and natural gas power plants. The environmental variable was expressed by the global warming potential of the systems, while Primary Energy Demand was the energy indicator and Levelised Cost of Energy (LCOE) served as a parameter of economic analysis.

Global warming and Primary Energy Demand were determined by attributional Life Cycle Assessment, which were applied according to a scope 'from cradle-to-gate' for the generation of 1.0 MWh of electricity.

The results obtained indicated that none of the assessed scenarios accumulated the best indexes in these dimensions. CO_2 and CH_4 are the major global warming impact precursors. S4 totals the positive effects of the implantation of the CaL process in coal- and gas- power plants. This is due to a decrease (62%) in total CO_2 emissions from coal and gas combustion. On the other hand, contributions to global warming derived from CH_4 emissions have increased significantly due to the energy requirements for CaL operation. This effect greatly compromised S4 performance, besides making the use of S3 unviable. CO_2 emissions from the $CaCO_3$ transport used in the CaL process also aided in neutralizing the gains that this technology provided.

As for Primary Energy Demand, S1 and S2 performed similarly and better than S3 and S4. The setbacks associated to S3 and S4 are due to crude oil consumption for diesel generation, for $CaCO_3$ transports, and raw natural gas, for CaL operation. This finding also undermines the application of that CCS alternative in natural gas thermoelectric plants.

The fact that the price of natural gas is twice as high as that of coal indicates that the use of CaL in natural gas power plants will result in higher costs than if the same system were coupled to coal-fired power plants.

The achievement of a combined verification of the three analysis dimensions, performed from normalized indicators, proved that S2 was the best option of the entire assessed series, due to the homogeneity of its indices. This indicates that the use of CaL in coal-fired power plants only appears as a viable alternative. The results obtained for S4 was unexpected. At the same time that their indices occupied the last positions in the energetic and economical dimensions, the simultaneous installation of CaL in coal-fired and gas thermoelectric plants was effective enough to capture and store CO_2 emissions, to the point of replacing this option in the set of possibilities considered in processes of decisions that occur within the scope of Brazilian energy planning.

This study met the original expectations of what was proposed, i.e., to help equate the existing dichotomy between the need for domestic electricity supply expansion and the reduction of global warming impacts that originate from electricity generation in Brazil, without, however, raising their associated costs, or imposing efficiency losses to the system.

Author Contributions: Conceptualization, C.C.S.M. and L.K.; Methodology, C.C.S.M.; Software, C.C.S.M.; Validation, C.C.S.M. and L.K.; Formal Analysis, C.C.S.M.; Investigation, C.C.S.M.; Data Curation, C.C.S.M.; Writing-Original Draft Preparation, C.C.S.M.; Writing-Review & Editing, L.K.; Supervision, L.K.; Project Administration, L.K.

Funding: This study was financed in part by the Coordenação de Aperfeiçoamento de Pessoal de Nível Superior-Brasil (CAPES) Finance Code 001.

Conflicts of Interest: The authors declare no conflict of interest.

References

1. International Panel on Climate Change (IPCC). *Climate Change 2014: Mitigation of Climate Change. Contribution of Working Group III to the Fifth Assessment Report of the Intergovernmental Panel on Climate Change*; Cambridge University Press: Cambridge, UK, 2014; 1435p, ISBN 978-1-107-05821-7.
2. Theo, W.L.; Lim, J.S.; Hashim, H.; Mustaffa, A.A.; Ho, W.S. Review of pre-combustion capture and ionic liquid in carbon capture and storage. *Appl. Energy* **2016**, *183*, 1633–1663. [CrossRef]
3. Kileber, S.; Parente, V. Diversifying the Brazilian electricity mix: Income level, the endowment effect, and governance capacity. *Renew. Sustain. Energy Rev.* **2015**, *49*, 1180–1189. [CrossRef]
4. Alam, M.M.; Murad, M.W.; Noman, A.H.M.; Ozturk, I. Relationships among carbon emissions, economic growth, energy consumption and population growth: Testing Environmental Kuznets Curve hypothesis for Brazil, China, India and Indonesia. *Ecol. Indic.* **2016**, *70*, 466–479. [CrossRef]
5. Turconi, R.; Boldrin, A.; Astrup, T. Life cycle assessment (LCA) of electricity generation technologies: Overview, comparability and limitations. *Renew. Sustain. Energy Rev.* **2013**, *28*, 555–565. [CrossRef]
6. Geller, M.T.B.; Meneses, A.A.D.M. Life Cycle Assessment of a Small Hydropower Plant in the Brazilian Amazon. *J. Sustain. Dev. Energy* **2016**, *4*, 379–391. [CrossRef]
7. Ministry of Mines and Energy (MME). Available online: http://www.mme.gov.br (accessed on 25 June 2018).
8. Koornneef, J.; Ramirez, A.; Van Harmelen, T.; Van Horssen, A.; Turkenburg, W.; Faaij, A. The impact of CO_2 capture in the power and heat sector on the emission of SO_2, NO_x, particulate matter, volatile organic compounds and NH 3 in the European Union. *Atmos. Environ.* **2010**, *44*, 1369–1385. [CrossRef]
9. Cuéllar-Franca, R.M.; Azapagic, A. Carbon Capture, storage and utilization technologies: A critical analysis and comparison of their life cycle environmental impacts. *J. CO_2 Util.* **2015**, *9*, 82–102. [CrossRef]
10. Anwar, M.N.; Fayyaz, A.; Sohail, N.F.; Khokhar, M.F.; Baqar, M.; Khan, W.D.; Rasool, K.; Rehan, M.; Nizami, A.S. CO_2 capture and storage: A way forward for sustainable environment. *J. Environ. Manag.* **2018**, *226*, 131–144. [CrossRef] [PubMed]
11. Cormos, C.C.; Petrescu, L. Evaluation of calcium looping as carbon capture option for combustion and gasification power plants. *Energy Procedia* **2013**, *51*, 154–160. [CrossRef]
12. Cormos, C.-C. Assessment of chemical absorption/adsorption for post-combustion CO_2 capture from Natural Gas Combined Cycle (NGCC) power plants. *Appl. Therm. Eng.* **2015**, *82*, 120–128. [CrossRef]
13. Blamey, J.; Anthony, E.J.; Wang, J.; Fennell, P.S. The calcium looping cycle for large-scale CO_2 capture. *Prog. Energy Combust. Sci.* **2010**, *36*, 260–279. [CrossRef]
14. Adanez, J.; Abad, A.; Garcia-Labiano, F.; Gayan, P.; De Diego, L.F. Progress in chemical-looping combustion and reforming technologies. *Prog. Energy Combust. Sci.* **2012**, *38*, 215–282. [CrossRef]
15. Biliyok, C.; Yeung, H. Evaluation of natural gas combined cycle power plant for post-combustion CO_2 capture integration. *Int. J. Greenh. Gas Control* **2013**, *19*, 396–405. [CrossRef]
16. Dieter, H.; Bidwe, A.R.; Varela-Duelli, G.; Charitos, A.; Hawthorne, C.; Scheffknecht, G. Development of the calcium looping CO_2 capture technology from lab to pilot scale at IFK, University of Stuttgart. *Fuel* **2014**, *127*, 23–37. [CrossRef]
17. Sreedhar, I.; Vaidhiswaran, R.; Kamani, B.M.; Venugopal, A. Process and engineering trends in membrane based carbon capture. *Renew. Sustain. Energy Rev.* **2017**, *68*, 659–684. [CrossRef]
18. Miccio, F.; Bendoni, R.; Piancastelli, A.; Medri, V.; Landi, E. Geopolymer composites for chemical looping combustion. *Fuel* **2018**, *225*, 436–442. [CrossRef]
19. Rochedo, P.R.R.; Costa, I.V.L.; Império, M.; Hoffmann, B.S.; Merschmann, P.R.D.C.; Oliveira, C.C.N.; Szklo, A.; Schaeffer, R. Carbon capture potential and costs in Brazil. *J. Clean. Prod.* **2016**, *131*, 280–295. [CrossRef]

20. Singh, B.; Strømman, A.H.; Hertwich, E. Life cycle assessment of natural gas combined cycle power plant with post-combustion carbon capture, transport and storage. *Int. J. Greenh. Gas Control* **2011**, *5*, 457–466. [CrossRef]

21. Korre, A.; Nie, Z.; Durucan, S. Life cycle modelling of fossil fuel power generation with post-combustion CO_2 capture. *Int. J. Greenh. Gas Control* **2010**, *4*, 289–300. [CrossRef]

22. Castelo Branco, D.A.; Moura, M.C.P.; Szklo, A.; Schaeffer, R. Emissions reduction potential from CO_2 capture: A life-cycle assessment of a Brazilian coal-fired power plant. *Energy Policy* **2013**, *61*, 1221–1235. [CrossRef]

23. National Electric Energy Agency (ANEEL). Generation Information Database (BIG). 2018. Available online: http://www2.aneel.gov.br/aplicacoes/capacidadebrasil/capacidadebrasil.cfm (accessed on 18 October 2018).

24. Barros, M.V.; Piekarski, C.M.; De Francisco, A.C. Carbon footprint of electricity generation in Brazil: An analysis of the 2016–2026 period. *Energies* **2018**, *11*, 1412. [CrossRef]

25. National Operator of the Electrical System (ONS). 2018. Available online: http://www.ons.org.br (accessed on 29 June 2018).

26. Energy Research Company (EPE). *Thermoelectric Power: Natural Gas, Biomass, Coal, Nuclear*; EPE: Rio de Janeiro, Brazil, 2016; 417p, ISBN 978-85-60025-05-3.

27. Union of Sugarcane Industry (UNICA). 2017. Available online: http://www.unica.com.br (accessed on 12 November 2017).

28. Guerra, J.P.; Cardoso, F.H.; Nogueira, A.; Kulay, L. Thermodynamic and environmental Analysis of Scaling up Cogeneration Units Driven by Sugarcane Biomass to Enhance Power Exports. *Energies* **2018**, *11*, 73. [CrossRef]

29. Energy Research Company (EPE). *Renewable Energy: Hydraulics, Biomass, Wind, Solar, Ocean*; EPE: Rio de Janeiro, Brazil, 2016; 452p, ISBN 978-85-60025-06-0.

30. Ministry of Mines and Energy (MME). Monthly Bulletin of Electrical Energy. 2018. Available online: http://www.mme.gov.br/web/guest/publicacoes-e-indicadores/boletins-de-energia (accessed on 10 October 2018).

31. Corsten, M.; Ramírez, A.; Shen, L.; Koornneef, J.; Faaij, A. Environmental impact assessment of CCS chains—Lessons learned and limitations from LCA literature. *Int. J. Greenh. Gas Control* **2013**, *13*, 59–71. [CrossRef]

32. Zhang, Z.; Huisingh, D. Carbon dioxide storage schemes: Technology, assessment and deployment. *J. Clean. Prod.* **2017**, *142*, 1055–1064. [CrossRef]

33. Fransson, E.; Detert, M. Process Integration of CO_2 Capture by Means of Calcium Looping Technology. Master's Thesis, Chalmers University of Technology, Gothenburg, Sweden, 2014.

34. Ruiz, M. Assessment of Calcium Looping as a Solution for CO_2 Capture in the Steel Production Process. Master's Thesis, University of Stuttgart, Stuttgart, Germany, 2015.

35. Jayarathna, C.K.; Mathisen, A.; Øi, L.E.; Tokheim, L.A. Process Simulation of Calcium Looping with Indirect Calciner Heat Transfer. In Proceedings of the 56th SIMS, Linköping, Sweden, 7–9 October 2015.

36. Shimizu, T.; Hirama, T.; Hosoda, H.; Kitano, K.; Inagaki, M.; Tejima, K. A twin fluid-bed reactor for removal of CO_2 from combustion processes. *Chem. Eng. Res. Des.* **1999**, *77*, 62–68. [CrossRef]

37. Valverde, J.M.; Sanchez-Jimenez, P.E.; Perez-Maqueda, L.A. Calcium-looping for post-combustion CO_2 capture. On the adverse effect of sorbent regeneration under CO_2. *Appl. Energy* **2014**, *126*, 161–171. [CrossRef]

38. Energy Research Company (EPE). *Brazilian Energy Balance 2018: Year 2017*; EPE: Rio de Janeiro, Brazil, 2018; 292p.

39. Vaz, C.E.M.; Maia, J.L.P.; Santos, W.G. *Tecnologia da Indústria do Gás Natural*, 1st ed.; Petrobras: Rio de Janeiro, Brazil, 2008; 440p, ISBN 9788521204213.

40. Vorrias, I.; Atsonios, K.; Nikolopoulos, A.; Nikolopoulos, N.; Grammelis, P.; Kakaras, E. Calcium looping for CO_2 capture from a lignite fired power plant. *Fuel* **2013**, *113*, 826–836. [CrossRef]

41. Martinez, I.; Murillo, R.; Grasa, G.; Abanades, J.C. Integration of a Ca-looping system for CO_2 capture in an existing power plant. *Energy Procedia* **2011**, *4*, 1699–1706. [CrossRef]

42. Hu, Y.; Ahn, H. Process integration of a Calcium-looping process with a natural gas combined cycle power plant for CO_2 capture and its improvement by exhaust gas recirculation. *Appl. Energy* **2017**, *187*, 480–488. [CrossRef]

43. Zbicinski, I.; Stavenuiter, J.; Kozlowska, B.; van de Coevering, H. *Product Design and Life Cycle Assessment*; The Baltic University Press: Nina Tryckeri, Uppsala, 2006; pp. 88–89.

44. International Organization for Standardization (ISO). *Environmental Management-Life Cycle Assessment-Principles and Framework*, 2nd ed.; ISO 14040:2006; ISO: Geneva, Switzerland, 2006.

45. International Organization for Standardization (ISO). *Environmental Management-Life Cycle Assessment-Requirements and Guidelines*, 1st ed.; ISO 14044:2006; ISO: Geneva, Switzerland, 2006.

46. Dones, R.; Bauer, C.; Röder, A. *Coal Energy Chain in USA. Final Report Ecoinvent N. 6-VI*; Swiss Centre for Life Cycle Inventories: Dübendorf, Switzerland, 2007.

47. Vasudevan, S.; Farooq, S.; Karimi, I.A.; Saeys, M.; Quah, M.C.G.; Agrawal, R. Energy penalty estimates for CO_2 capture: Comparison between fuel types and capture-combustion modes. *Energy* **2016**, *103*, 709–714. [CrossRef]

48. National Confederation of Transport (CNT). Available online: http://www.cnt.org.br/Paginas/atlas-do-transporte (accessed on 20 August 2018).

49. Goedkoop, M.; Heijungs, R.; Huijbregts, M.; De Schryver, A.; Struijs, J.; van Zelm, R. Description of the ReCiPe Methodology for Life Assessment Impact Assessment. 2013. Available online: http://www.lciarecipe.net (accessed on 17 May 2018).

50. Frischknecht, R.; Jungbluth, N.; Althaus, H.-J.; Bauer, C.; Doka, G.; Dones, R.; Hischier, R.; Hellweg, S.; Humbert, S.; Köllner, T.; et al. *Implementation of Life Cycle Impact Assessment Methods: Data v2.0. Ecoinvent report No. 3*; Swiss Centre for Life Cycle Inventories: Dübendorf, Switzerland, 2007.

51. Romeiro, D.L.; Almeida, E.; Losekann, L. The choice of dispatchable versus intermittent electricity generating technologies and the Brazilian case. In Proceedings of the 5th Latin American Energy Economics Meeting, Medellín, Colombia, 15–18 March 2015.

52. International Energy Agency (IEA). *Projected Costs of Generating Electricity*; OECD Publishing: Paris, Italy, 2015; 213p.

53. Energy Research Company (EPE). *Ten Year Energy Expansion Plan 2026*; MME/EPE: Brasília, Brazil, 2017; 271p.

54. Mantripragada, H.C.; Rubin, E.S. Calcium Looping Cycle for CO_2 Capture: Performance, Cost and Feasibility Analysis. *Energy Procedia* **2014**, *63*, 2199–2206. [CrossRef]

Article

Analysis of Carbon Storage and Its Contributing Factors—A Case Study in the Loess Plateau (China)

Gaohuan Liu [1] and Zhonghe Zhao [1,2,*]

[1] State Key Laboratory of Resources and Environmental Information System, Institute of Geographic Sciences and Natural Resources Research, Chinese Academy of Sciences, Beijing 100101, China; liugh@lreis.ac.cn
[2] University of Chinese Academy of Sciences, Beijing 100049, China
* Correspondence: zhaozh.16b@igsnrr.ac.cn; Tel.: +86-010-6488-9316

Received: 14 May 2018; Accepted: 12 June 2018; Published: 19 June 2018

Abstract: The Chinese Loess Plateau is an ecologically fragile and sensitive area. The carbon storage dynamics in this region and the contributions from land use/land cover change (LUCC) and carbon density from 2000 to 2010 were analyzed in this paper. Normalized difference vegetation index (NDVI), biomass and soil carbon data in 2000 were used for regression analysis of biomass and soil carbon, and an inversion analysis was used to estimate biomass and soil carbon in 2005 and 2010. Quadrat data, including aboveground biomass and soil organic carbon, were used to calibrate the model output. Carbon storage and sequestration were calculated by the InVEST toolset with four carbon pools, including aboveground biomass, belowground biomass, dead wood and soil carbon. The results showed that carbon storage increased steadily from 2000 to 2010, increasing by 0.260 billion tons, and that woodland area increased and arable land decreased; the overall trend in land cover improved, but the improvement was not pronounced. Carbon storage in the Loess Plateau was correlated with geographical factors. We found that when assuming a constant carbon density, carbon storage decreased, accounting for −1% of the carbon storage dynamics. When assuming no land conversion, carbon storage increased, accounting for 101% of the carbon storage dynamics.

Keywords: carbon storage; the Loess Plateau; InVEST; carbon density; normalized difference vegetation index (NDVI)

---

1. Introduction

Land use/land cover change (LUCC) has prominent impacts on carbon storage and sequestration dynamics in ecosystems [1–5]. LUCC influences carbon storage in two ways: via land conversion and land modification. Several studies have focused on land conversion, whereas studies on land modification are scarce. LUCC has been a popular research topic in resource-related and environmental disciplines and is the core program for three international organizations, namely, the International Geosphere-Biosphere Programme (IGBP), the International Human Dimensions Programme (IHDP), and the World Climate Research Program (WCRP) [1–3]. If we cannot quickly and accurately understand LUCC information, we cannot correctly evaluate LUCC effects, such as soil erosion, on the ecological environment, making it impossible to accurately estimate the capacity of the ecosystem carbon cycle [4]. China has adopted a variety of ecological restoration projects to improve the ecological environment; these projects include the Natural Forest Protection Project and the Grain for Green Program (GGP). The GGP is the largest reforestation project in China [5]. According to domestic and foreign scholars, since the implementation of this project, which returned farmland to forest from 1990 to 2010, China has had the largest afforestation area in the world [6]. The Kyoto Protocol went into effect on 16 February 2005, and clearly stated that humans could increase carbon sequestration through the effective management of LUCC, for example, by reducing deforestation, increasing afforestation and effectively managing terrestrial ecosystem services [7].

Global carbon change has become a popular research topic [8], with many scholars having estimated carbon storage and sequestration, and estimation accuracy has gradually improved [9,10]. Most carbon is stored in the soil, terrestrial vegetation and atmosphere; approximately 1400–1500 Gt of carbon exists in the soil, approximately 500 Gt of carbon exists in terrestrial vegetation, and approximately 750 Gt of carbon exists in the atmosphere [11].

The ecological system of the Loess Plateau is fragile and sensitive, in recent decades, China has implemented several major ecological construction projects, including GGP, Ecological Construction Program, Sand Source Improvement Project, Small Watershed-Based Management Program, etc. The implementation of the abovementioned national level program greatly changed the land use and land cover. These changes have induced a big change in carbon storage. With the continuous efforts in ecological protections, the crop land reduced and natural vegetation area increased, especially the forest area increased which immensely improved ecological quality, and the carbon sequestration capacity increased. Study of carbon storage change induced mainly by LUCC and other factors can enhance the understanding of the process of carbon storage change. Thus, developing a carbon sink in Loess Plateau based on the previous development of carbon sequestration in forests, grasslands and wetlands to form a multilevel and multidimensional carbon trading market. It also can enhance economic and social development in Loess Plateau to achieve green development while sharing development goals. This study can therefore provide references for making decisions about ecological conservation and regional development.

2. Materials and Methods

2.1. Study Area

The Loess Plateau (34~40° N, 103~114° E) is the largest loess deposition area in the world with an elevation of 1000~1500 m. It spans seven provinces, namely, Qinghai, Gansu, Ningxia, Inner Mongolia, Shaanxi, Shanxi and Henan, 50 prefectural-level divisions and 284 county-level divisions [12]. The Loess Plateau is in a semiarid, subhumid area that is has a continental monsoon climate; the annual average temperature is 6~14 °C, and the average rainfall is 200~700 mm. The Loess Plateau has attracted the attention of domestic and foreign scholars for many years due to the complex relationships between humans and the Earth, the deteriorating ecological environment and serious soil erosion. To date, research on the ecosystem services of the Loess Plateau has been focused at the river-basin and single-forest (mostly locust forest) scales, and functional analyses of the ecosystem services in the entire region of the Loess Plateau are lacking [13–16].

Large-scale soil and water conservation projects have been implemented in Loess Plateau since the 1970s. China also implemented a restoration project to return farmland to forest and grassland after 1999, and the Loess Plateau was the regional focus of the implementation. Over the past few decades, many artificial forests have been planted in the area, which account for 59.8% of the total forest area in the Loess Plateau. Locust is the primary tree species that has been planted in Loess Plateau because it has a rapidly growing, drought-resistant, infertile root system. Locusts account for 90% of the artificial forest area.

2.2. Data Sources

2.2.1. LUCC Data

Land use in China at the 1:100,000 scale from remote sensing monitoring data sets represents the land use monitoring data product with the highest precision in China. In this study, three years were considered: 2000, 2005 and 2010. Remote sensing images from Landsat TM/ETM were the main data source and were analyzed through artificial visual interpretation. Six primary types of land use data were used, namely, data for cultivated land, forestland, grassland, water area, building area (urban land and rural land) and unused land, and 25 secondary types were used. During the process of analyzing carbon storage, five primary land use types were used: cultivated land, forestland, grassland, building land and unused land. Mountain paddy fields and hilly paddy fields, plain paddy fields and 18 other land use types were also analyzed. Water, sand and other area types were not included. The data set was provided by the Data Center for Resources and Environmental Sciences, Chinese Academy of Sciences (RESDC) (http://www.resdc.cn).

2.2.2. NDVI

The NDVI data were from MOD13Q1, which has a temporal resolution of 16 days and a spatial resolution of 250 m. The current study used the average NDVI during the growing season from June to August in 2000, 2005, and 2010. The data set was provided by the International Scientific & Technical Data Mirror Site, Computer Network Information Center, Chinese Academy of Sciences (http://www.gscloud.cn).

2.2.3. Digital Elevation Model (DEM)

The DEM data were from SRTMDEMUTM, which has a spatial resolution of 90 m. The data set was provided by the International Scientific & Technical Data Mirror Site, Computer Network Information Center, Chinese Academy of Sciences (http://www.gscloud.cn).

2.2.4. Slope

The DEM data were processed using the slope function in ArcGIS Spatial Analyst. The slope data at 90-m spatial resolution were obtained from the slope calculated from the DEM data in the study area.

2.2.5. Precision-Test Data

Precision-test data were acquired from MYD17A2 and MOD17A2, which each have a spatial resolution of 1000 m and a temporal resolution of 8 days. The data set was provided by the International Scientific & Technical Data Mirror Site, Computer Network Information Center, Chinese Academy of Sciences (http://www.gscloud.cn). In addition, survey data, 146 soil samples, 12 tree plots, and three grassland surveys were used.

2.2.6. Carbon Density Parameters of 2000

Biomass carbon density data were acquired from the Carbon Dioxide Information Analysis Center (CDIAC) [17] and were based on the biomass carbon density from the Intergovernmental Panel on Climate Change (IPCC). Then, 1 km global biological carbon density raster data were generated from the land use and vegetation flora data. Soil carbon density data were acquired from the Joint Research Center (JRC) of the European Union in Brussels along with surface soil organic carbon (0–30 cm) data [18]. Humus carbon data and root-to-shoot ratios were determined from the IPCC 2006 national greenhouse gas emissions [19].

2.3. Research Methods

We used the data set in 2000 to construct two regression models: an NDVI and biomass regression model and a biomass and soil organic carbon regression model. Then, the biomass and soil organic carbon data in 2005 and 2010 were modeled based on model inversion. Biomass was converted into aboveground biomass and underground biomass based on the root-to-shoot ratio. Because the humus carbon pool is small and stable, the humus carbon density data in this paper represent the data collected in 2000. Then, the land cover and carbon density parameters were input into Integrated Valuation of Ecosystem Services and Trade (InVEST) to determine the carbon storage in 2005 and 2010 in the Loess Plateau.

(1) The key carbon library parameter inversion

GIS spatial analysis was used to acquire the mean NDVI, biomass and soil organic carbon values of the 18 land use types. Then, a statistical analysis was conducted in SPSS 21.0. NDVI, biomass, and soil organic carbon (SOC) exhibit a normal distribution. NDVI and biomass were significantly related to SOC (alpha = 0.05), yielding correlation coefficients of R^2 = 0.883 and R^2 = 0.716, respectively. Biomass was significantly related to SOC (alpha = 0.05), with a correlation coefficient of R^2 = 0.900, according to the regression analysis. In accordance with the requirement of a strong correlation, a high biomass-NDVI quadratic curve model and an SOC-biomass quadratic curve model were constructed (Figure 1). The equation was derived from an analysis of variance and a regression coefficient *t*-test ($p < 0.01$). The regression equation was based on NDVI, biomass and SOC as well as the inversion of the key parameters measured in 2005 and 2010, such as biomass and SOC.

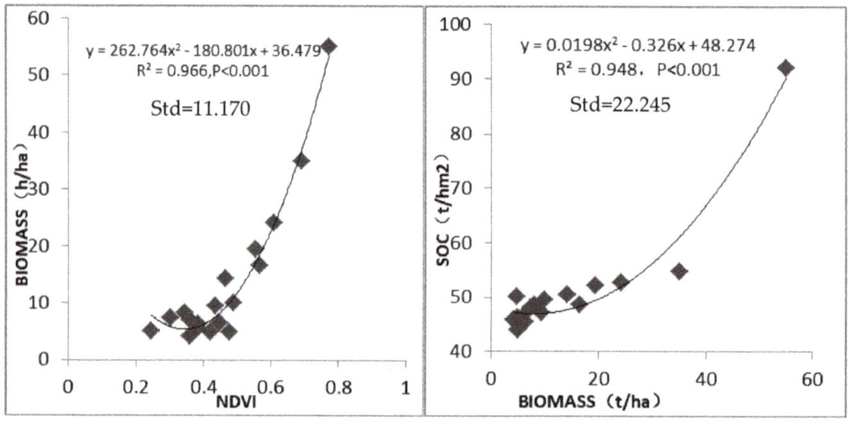

Figure 1. Biomass-NDVI and SOC-NDVI regression analyses.

(2) Ecosystem carbon storage simulation

InVEST, developed at Stanford University, is a comprehensive ecosystem service evaluation model that can assess various ecosystem services, provide comprehensive analysis for planning ecological restoration, pay ecosystem services (PES), and assess developmental effects and space permits. Cooperation risk management plays an increasingly important role in promoting biodiversity conservation, and the coordinated development of human well-being is of great significance [20]. Land cover and biomass data as well as soil carbon and humus carbon data collected in the study area in 2000, 2005, and 2010 were used as InVEST input data to simulate carbon storage during the corresponding years. Then, the behavior and spatial differentiation of carbon storage in Loess Plateau from 2000–2010 was analyzed. The carbon computation formula in the carbon module of the InVEST model is described as follows:

$$C_{zone} = \sum C_i \times A_i \tag{1}$$

$$C = C_{above} + C_{below} + C_{dead} + C_{soil} \tag{2}$$

where C_{zone} is the carbon of the zone, C is the total carbon, A_i is the area of class I, C_i is the carbon density of class I (t C/ha), C_{above} represents aboveground biomass, C_{below} represents root biomass, C_{dead} represents humus carbon, and C_{soil} represents soil organic carbon.

(3) Carbon storage function change analysis of contributing factors

In general, changes in land use can be divided into changes in land cover type (land conversion) and changes in the internal quality of a specific land cover type (land modification). Accordingly, changes in the regional carbon storage function involve changes in land cover type and carbon density and the contributions of different factors that affect the carbon storage function. The model of this process is described as follows:

(1) Changes in regional carbon storage function based on two carbon available libraries using the ΔC characteristic:

$$\Delta C = \sum_{i=1}^{n} \left(A_{i2} D_{i2} - A_{i1} D_{i1} \right) \tag{3}$$

where ΔC is regional carbon storage change; A_{i1} and A_{i2} are the areas of class i before and after the change, respectively; and D_{i1} and D_{i2} are the carbon densities of class i before and after the change, respectively.

(2) Assuming that the carbon density of each land use/land cover type is constant and that the carbon storage changes are caused only by changes in land use/land cover, the carbon storage change can be represented as follows:

$$\Delta C_1 = \sum_{i=1}^{n} \left(A_{i2} - A_{i1} \right) D_{i1} \tag{4}$$

where ΔC_1 is the carbon storage change due only to changes in land use/land cover.

(3) Assuming that land use/land cover is constant and that the carbon storage changes are caused only by carbon density changes, the carbon storage change can be represented as follows:

$$\Delta C_d = \sum_{i=1}^{n} A_{i1} \left(D_{i2} - D_{i1} \right) \tag{5}$$

where ΔC_d is the carbon storage change due only to C carbon density changes.

(4) The contribution rate of the change in land use/land cover types to carbon storage (R_l) and the contribution rate of the change in carbon density to carbon storage (R_d) can be calculated using the following formulas:

$$R_l = \Delta C_l / (\Delta C_l + \Delta C_d) \times 100\% \tag{6}$$

$$R_d = \Delta C_d / (\Delta C_l + \Delta C_d) \times 100\% \tag{7}$$

where R_l is the contribution rate of the change in land use/land cover types to carbon storage and R_d is the contribution rate of the change in carbon density to carbon storage.

2.4. Precision Testing and Sample Analysis

Remote sensing images and sampling point data were used to verify the accuracy of the models. The entire Loess Plateau area and the individual land areas and points were tested to determine if this method was reliable.

The area of the Loess Plateau is vast, and the terrain is complex. Thus, carbon storage in the Loess Plateau was simulated and analyzed over a large scale, and the accuracy of the simulation was assessed over the entire surface area and at a typical area. The accuracy over the surface area in this study was verified by the MODIS MOD17A2 and MYD17A2 data products for 2010. The vegetation biomass data and the soil organic matter data were used to verify the accuracy over the typical area.

The MODIS total primary productivity products, MYD17A2 and MOD17A2, depict the cumulative biological (mainly green plants) organic carbon fixed through the photosynthetic pathway, which is same as the biomass (total of aboveground biomass and belowground biomass) parameters in the InVEST model. In this study, the total primary productivity of the Loess Plateau in 2010 was 8.61×10^8 t C, which was calculated by the carbon parameter inversion. The total primary productivity estimated from the MYD17A2 data product in the Loess Plateau in 2010 was 8.16×10^8 t C; the relative error was 5.2%. The total productivity estimated from the MOD17A2 data product in the Loess Plateau in 2010 was 8.95×10^8 t C; the relative error was -3.9%. Therefore, the simulation results of this study can be used for further dynamic change analysis.

The typical area (Figure 2) considered in this study is in Shilou County, Shanxi Province, which is an experimental area where farmland was converted to forest (36°56′ N, 110°46′ E). The study area is near the east bank of the Yellow River. The terrain is high in the east and low in the west, and there are rolling mountains, crisscrossed gullies, and broken terrain. The study area is located along the middle reaches of the Yellow River, where soil erosion is very serious. The landform type is typical of the loess hilly and gully region of the Loess Plateau, and the total area of the district is 2.02 km^2. The terrain is undulating, with a relative height difference of more than 200 m. Most trenches have been cut to bedrock. The soil in this area mainly comprises yellow and gray-brown soils, which are characterized as silty clay loam, and the parent material is loess.

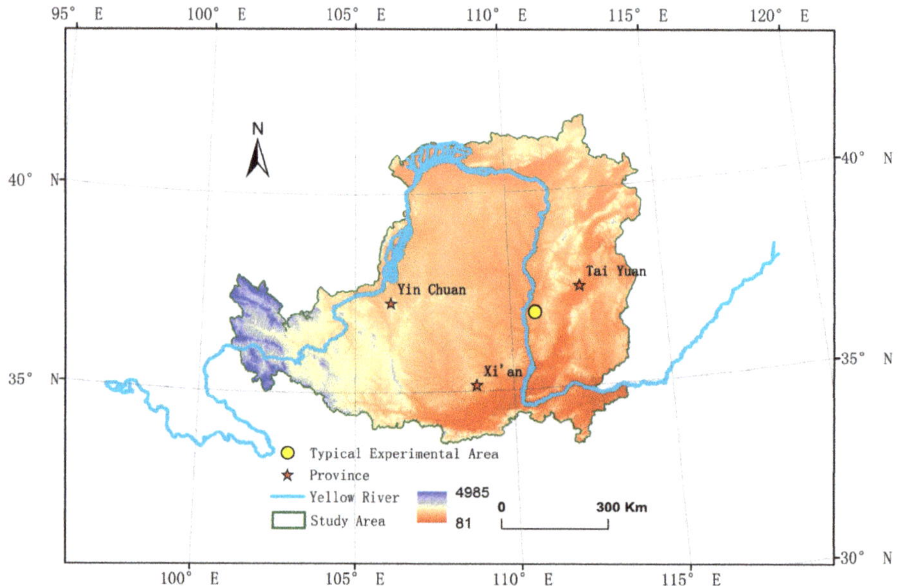

Figure 2. Location of the typical experimental area.

The area has a warm temperate continental monsoon climate. The meteorological data of the area for 1981–2010 are as follows: mean temperature of 9.8 °C; average monthly rainfall of 38.7 mm mainly concentrated from July to September; and monthly average humidity of 55.3%. From 2011 to 2015, the average monthly temperature was 10.7 °C; the average monthly rainfall was 45.6 mm, which was also mostly concentrated from July to September, accounting for more than 60% of the annual rainfall; and the average monthly humidity was 53%.

The main land cover/land use types in the typical area are locust forest, hill dryland, low-coverage grassland, and low human interference. In recent years, the Hong Kong Forces Planted Charitable Foundation improved the ecological environment of the typical experimental area by continuously growing artificial forest over 130 hectares of the area. Locusts were the primary trees planted, accounting for more than 98% of the tree planting area, followed by *Armeniaca sibirica*, *Sophora japonica*, and *Rhus typhina*. The main plantation model involved digging equal pits (2 m long, 50 cm wide, and 50 cm deep) in a "triangle"-shaped arrangement, with the center of the pit 3 m from the column and 1.5 m from the row, and there were approximately 2250 pits per ha.

The survival rate of the plants in this typical area is greater than 80%, which is much higher than the local afforestation survival rate. The forest development is good, which is conducive to improving soil conditions and conserving water. The area has been cultivated for more than 30 years, and the crop plants are grown once per year. The main crops are corn and millet. *Cirsium setosum, Artemisia annua, Setaria viridis* and other plants are the more common in weeds in the area.

Twelve forest plots were randomly selected in the typical experimental area, and the size of each plot was 10 × 10 m. The survey indicators mainly included planting density, tree height, branch height, basal diameter, diameter at breast height (DBH), crown width and biomass. The heights of the trees were measured using a TruPulse200 rangefinder (Laser Technology, Inc., Centennial, CO, USA), and the diameter at breast height was measured in feet. The crown width was measured using a steel tape; then, the biomass was calculated using the relationships between *Robinia pseudoacacia* growth parameters (tree height, DBH) and biomass. In addition, forest-herb samples were randomly selected under the trees in each locust sample. Each 1 × 1 m forest-herb sample was selected randomly. After the grass was mowed, cleaned, and dried, the ground biomass was weighed to the nearest 0.01 g (Guangzheng YP-B electronic scale, Shanghai, China), and its carbon content was converted by a ratio of 0.47. In the typical experimental area, we randomly selected grassland samples, and the size and treatment methods were the same as those applied to the forest-herb samples.

The soil in the typical experimental area was sampled in May 2016. Four layers of soil were surveyed at each sample point: 0–10 cm, 10–20 cm, 20–40 cm, and 40–60 cm. After sampling, the samples were taken to the laboratory for physical and chemical analyses. The SOC and bulk density were analyzed. There was a total of 140 forest soil samples, 12 grassland samples, and 12 cultivated land samples, yielding an overall total of 164 samples. The SOC was measured by potassium dichromate oxidation spectrophotometry (HJ615-2011).

(1)   Calculation of the carbon density of the forest

The planting distance and seedling growth of the acacia forest are the same, which is consistent with individual nutrition space. Outside these growing conditions, reproductive success is difficult without human interference. Therefore, when the survey found sparsely scattered biomass in the forest, the forest aboveground biomass depended on the density of the living trees, the biomass per tree and the biomass of the understory plants. Locust forest biomass included the biomass of five parts: trunks, branches, leaves, bark and roots. The locust biomass was calculated using the allometric equations published by the SFA (State Forestry Administration) [21] as follows:

$$W_T = 0.02583 \times \left(D^2 H\right)^{0.6841} \tag{8}$$

$$W_B = 0.00464 \times D^{3.2181} \tag{9}$$

$$W_L = 0.02340 \times D^{1.9277} \tag{10}$$

$$W_p = 0.00763 \times \left(D^2 H\right)^{0.0447} \tag{11}$$

$$W_R = 0.01779 \times D^{2.6148} \tag{12}$$

where W_T, W_B, W_L, W_P, *and* W_R represent the biomass of trunks, branches, leaves, bark and roots per tree (kg), respectively; D represents the breast diameter (cm); and H represents the height of the tree (m).

The biomass of each part was calculated and then multiplied by the carbon content coefficient and the corresponding stand density to obtain the aboveground carbon density. According to the survey, the stand density was 330 trees per ha, and the carbon contents coefficient of the trunks, branches, leaves, bark and roots were 0.497, 0.481, 0.465, 0.458 and 0.460, respectively.

(2) Calculation of the carbon density of understory plants or grassland

The chemical substance return rate and the biomass reduction rate of understory plants are higher than those of the upper layer of the forest, and the effect of ecosystem circulation cannot be underestimated. The carbon storage of the forest-herb under the trees was calculated using the aboveground biomass, which was then converted to carbon content. In addition, the method used to calculate the carbon content of the grassland was the same as that used to calculate the carbon content of the understory plants:

$$C_g = BIO_g \times T_g \tag{13}$$

where C_g represents the carbon content of the forest-herb under the trees (t C/ha), BIO_g represents the dry weight of the aboveground biomass per ha (t), and T_g represents the conversion of carbon-containing herbs (%). In addition, grassland carbon was calculated in the same manner as the carbon of the understory plants, and the carbon-containing herb conversion [19].

According to the allometric growth equation, the carbon density of the locust trees in the typical experimental area was 12.54 t C/ha, and the carbon density of understory plants was 0.89 t C/ha. Thus, the carbon density of the forest was 13.43 t C/ha, whereas the carbon density in the model was 13.07 t C/ha in 2010, representing an accuracy of 97.3%. The amount of aboveground biomass per year is directly related to climate, rainfall, light and other parameters. The aboveground biomass fluctuates over time and is not a steadily increasing or fixed value. However, woodlands are different from cultivated lands, as trees continue to grow and the biomass continues to increase. The accuracy is within the allowable range of error, indicating that the model could be used for further simulation and analysis.

In the survey, the grassland carbon density was 1.38 t C/ha. The corresponding density in the model was 1.54 t C/ha due to drought in the typical experimental area that year and the influences of vegetation growth, thus resulting in the low accuracy of 88.4%. Because of the special climate conditions in the typical experimental area during that year, the model precision is adaptable when considering the Loess Plateau.

(3) Calculation of soil carbon density

Regional differences in soil carbon density are obvious, but the values are largely stable over time. Changes in soil carbon density are closely related to land use/land cover. The soil carbon density calculation method is as follows:

$$SOC_i = 0.1 \times C_i \times D_i \times E_i \times (1 - G)/100 \tag{14}$$

where SOC_i represents the soil carbon storage (Mg/ha), C_i represents the soil carbon organic content of the i-th layer (g/kg), D_i represents the soil bulk density (g/cm^3), E_i represents the thickness of the i-th layer of soil (cm); and G_i represents the volume content, which is the gravel in the soil greater than 2 mm in the i-th layer (%). As the loess, hilly and gully region is developed from loess parental material, the gravel content is low, so $G_i = 0.5$.

Based on the calculation of soil carbon density of the 164 soil samples, the soil carbon density of the locust forest was 51.66 t C/ha. The coverage of weeds in the typical experimental area was low, and the carbon density was 62.18 t C/ha. The soil carbon density of the hilly dryland was 56.17 t C/ha. According to the regression curve of the model, the soil carbon density of the shrubwood was 53.85 t C/ha and 51.66 t C/ha in 2005 and 2010, respectively, and the accuracy was 95.76% and 97.06%, respectively. The soil carbon density of the hilly dryland was 54.91 t C/ha and 58.13 t C/ha in 2005 and 2010, respectively, and the accuracy was 97.76% and 96.51%, respectively. The soil carbon density of the low-coverage grassland was 62.51 t C/ha and 62.94 t C/ha, and the accuracy was 99.47% and 98.78%. Therefore, the regression curve and model could be used for further analysis.

3. Results

The land use/land cover data and the corresponding types of carbon densities were input into the InVEST model, and carbon storage and sequestration were simulated in the Loess Plateau for the three years of 2000, 2005, and 2010. In addition, the contribution of land use change to carbon sequestration, the contribution of carbon footprint change to carbon storage, and the relationships between carbon storage change and geographical factors (elevation, slope) were analyzed.

3.1. Carbon Storage Patterns

From 2000 to 2010, carbon storage increased steadily in the Loess Plateau. Carbon storage in 2000, 2005 and 2010 was 3.95, 3.96, and 4.21 billion tons, respectively. Total carbon storage increased by 0.26 billion tons, and carbon storage showed steady growth. The land cover types distributed from the northwest to the southeast were grassland, forest and cultivated land. Vegetation coverage increased gradually across the Loess Plateau. Due to measures such as the return of farmland to forest, afforestation, the planting of grass, the construction of horizontal terraces, the implementation of water conservancy projects and the comprehensive management of small watersheds in Loess Plateau, the vegetation area of the Loess Plateau is gradually increasing, and vegetation growth is improving. The distribution pattern of carbon storage is consistent with that of natural zonation, and carbon storage in the ecosystem gradually increased from the northwest to the southeast (Figure 3).

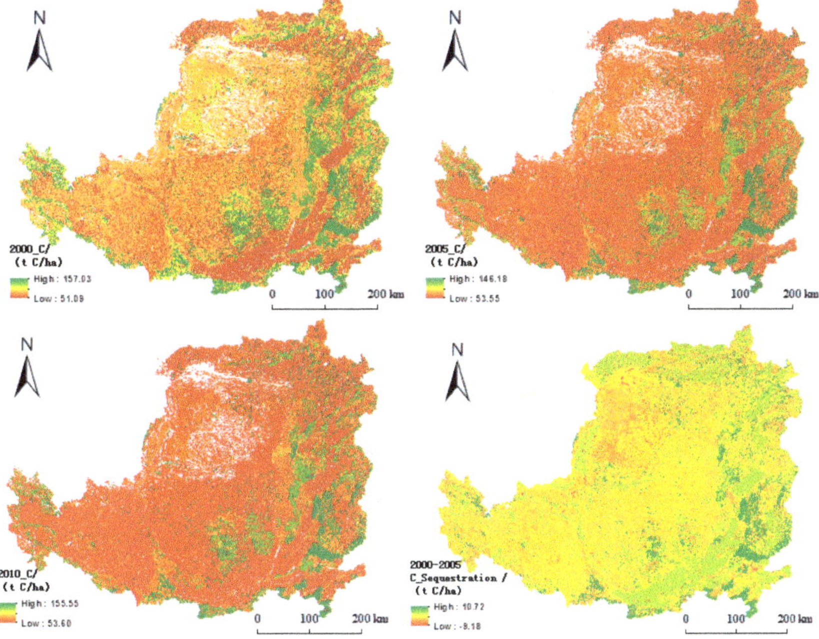

Figure 3. *Cont.*

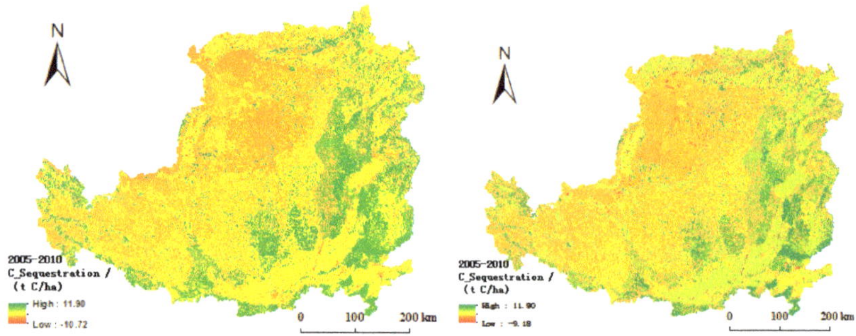

Figure 3. Temporal and spatial patterns of carbon storage in Loess Plateau from 2000 to 2010.

3.2. The Contribution of Changes in Land Use Type to Carbon Storage

The climate of the Loess Plateau shifts from warm subhumid to semiarid climate to arid from the southeast to the northwest. The natural vegetation corresponds to the climate type and is forest-grassland, grassland, and sand-grassland. The difference in the terrain of the land use type and vegetation coverage is obvious. The growth rate of vegetation is significant, and reconstruction and conservation occurred in the major areas of vegetation restoration on the sloping fields between 2000 and 2005.

The effect of LUCC on vegetation coverage increased, and the growth rate of vegetation coverage in the LUCC area was significantly higher than that in the unchanged area. From 2000 to 2010, the growth rates of the areas where farmland was returned to forest and grassland were especially prominent. Grassland areas first decreased and then increased. The area with high grassland coverage increased to approximately 1153 km^2, and the area with moderate or low grassland coverage decreased to 1789 km^2. The areas of all types of forest increased markedly. The forest area increased by 503 km^2, the shrubwood area increased by 449 km^2, the sparse-woods area increased by 134 km^2, and other woodlands increased by 1723 km^2. The area with zero carbon increased by 1159 km^2, representing a percentage increase of 0.2%. The areas of rivers and canals, reservoir pits, other construction lands, desert, bare lands and bare rock mass all increased. The areas of lakes, bottomland and the Gobi Desert all decreased. The areas of urban and rural land increased, the saline-alkali soil area decreased, and the marshland area increased. The total area of cultivated land, such as mountain paddy fields, hill paddy fields, plain paddy fields, mountain dryland, hilly dryland, plain dryland and dryland (>25 slope), decreased from 216,476 km^2 to 211,736 km^2 (Figure 4).

Assuming that the carbon density is constant, the carbon storage changes caused only by the changes in land use/land cover type were considered. From 2000 to 2010, the carbon storage in Loess Plateau decreased by 0.3 million tons, accounting for −1% of the carbon storage dynamics (Figure 5).

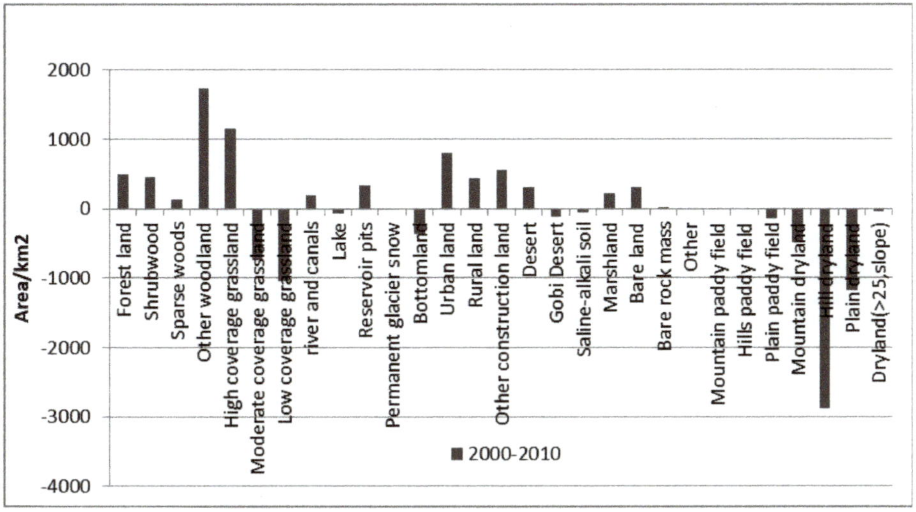

Figure 4. Land use structure changes from 2000 to 2010.

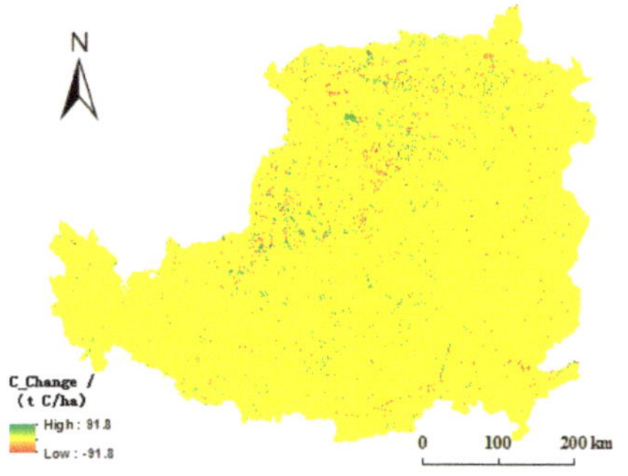

Figure 5. Changes in carbon storage due to LUCC from 2000 to 2010.

3.3. The Contribution of Carbon Density to Carbon Storage

From 2000 to 2010, the carbon densities of the areas with high carbon densities increased; these areas included forestland, shrubwood, sparse woods, and cultivated land. The carbon densities of most types of land increased. The carbon density slightly decreased in only four types of land, namely, high-coverage grassland, moderate-coverage grassland, low-coverage grassland and mountain paddy fields. The carbon densities of sparse woods and hilly drylands first decreased and then increased, and the overall trend was one of increase. The carbon density of the saline-alkali soil first increased and then decreased, and the overall trend was one of increase. The carbon densities of the other land use types, such as forestland, shrubwood, other woodlands, urban land, rural land, marshland, hilly paddy fields, plain paddy fields, mountain dryland, plain dryland and dryland (>15 slope), steadily increased (Table 1).

Table 1. The carbon density of different land use types and changes in carbon density during 2000–2010.

Land Type	2000/(t/ha)	2005/(t/ha)	2010/(t/ha)	2010–2000/(t/ha)	2005–2000/(t/ha)	2010–2005/(t/ha)
Forest land	91.79	107.17	118.99	15.38	11.83	27.2
Shrubwood	86.91	90.77	104.04	3.86	13.27	17.13
Sparse woods	73.68	71.66	81.26	−2.02	9.6	7.58
Other woodland	58.79	56.73	60.18	−2.06	3.45	1.39
High-coverage grassland	74.85	67.76	71.07	−7.08	3.31	−3.77
Moderate-coverage grassland	66.80	62.94	64.41	−3.86	1.47	−2.39
Low-coverage grassland	65.42	62.51	62.94	−2.91	0.43	−2.48
Urban land	51.38	53.55	53.60	2.17	0.06	2.23
Rural land	52.86	58.03	60.27	5.18	2.23	7.41
Saline-alkali soil	52.30	54.69	53.77	2.39	−0.92	1.47
Marshland	51.09	54.48	55.56	3.39	1.08	4.47
Mountain paddy field	157.03	146.18	155.55	−10.85	9.37	−1.48
Hills paddy field	59.00	62.92	75.61	3.92	12.7	16.61
Plain paddy field	64.92	69.02	71.79	4.1	2.77	6.87
Mountain dryland	60.64	63.05	73.42	12.78	2.41	10.38
Hill dryland	55.53	54.91	58.13	2.60	−0.62	3.22
Plain dryland	52.86	58.48	61.72	8.86	5.62	3.24
Dryland (>25 slope)	66.32	73.44	86.61	20.29	7.12	13.17

In total, the average carbon density steadily increased from 69.01 t/ha in 2000 to 70.46 t/ha in 2005 and to 76.05 in 2010, and the average increase was 7.04 t/ha. Assuming no land conversion, the carbon storage changes due only to the changes in carbon density were considered. The Loess Plateau carbon storage increased by 26.2 million tons, accounting for 100.1% of the carbon storage dynamics (Figure 6).

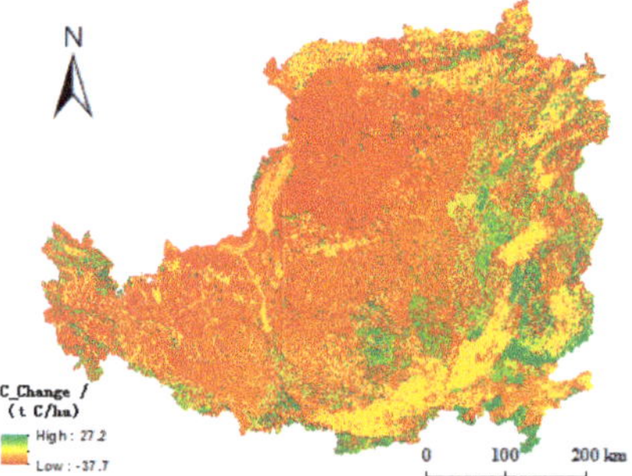

Figure 6. Changes in carbon storage due to changes in carbon density from 2000 to 2010

3.4. The Relationships between Carbon Storage and Geographical Factors

A DEM, a discrete digital representation of the surface topography of the earth [11], can be used to extract terrain factors, such as slope, aspect and slope position, and is widely used in geoanalysis [22]. In this study, the differences in carbon storage among different slope conditions were analyzed based on a DEM. Slope is a terrain factor derived from the DEM and not only represents the degree of tilt but also affects the redistribution of materials, such as soil, water, heat, and nutrients [23].

Carbon storage significantly differed among different slopes (Table 2). The vegetation on the Loess Plateau was mainly distributed in the areas with slopes of 0~26°. The vegetation area decreased with increasing slope. The 0°, flat region along both sides of the Yellow River is mostly cultivated land, and the Mausu Desert is in the northwest region of the Loess Plateau. As the slope increased, the land use type transitioned to woodland and grassland, the vegetation coverage improved, human disturbance decreased, and the average carbon density steadily increased (Figure 7).

Table 2. Carbon density of areas with different slopes in the Loess Plateau.

Slope (°)	2000 (t/ha)	2005 (t/ha)	2010 (t/ha)
0	61.70	62.37	65.14
1	63.79	63.68	67.29
2	65.52	65.41	69.62
3	67.16	67.21	71.84
4	68.92	69.29	74.39
5	70.97	71.71	77.28
6	72.87	74.17	80.28
7	74.28	75.85	82.34
8	75.27	77.05	83.70
9	76.16	77.85	84.67
10	77.27	79.51	86.61
11	77.47	79.72	86.81
12	77.74	79.97	87.13
13	78.58	81.33	88.75
14	78.59	81.33	88.79
15	78.26	80.93	88.33
16	78.82	81.74	89.05
17	79.20	82.53	89.92
18	79.26	82.45	90.40
19	79.19	82.08	89.54
20	79.51	83.24	91.18
21	80.08	84.29	92.62
22	79.41	85.50	93.34
23	81.03	83.19	92.12
24	86.14	94.03	103.02
25	83.32	87.47	95.03
26	91.79	107.17	118.99

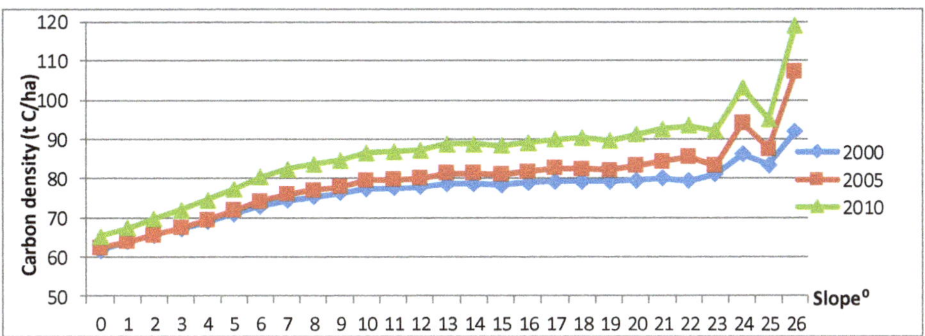

Figure 7. Relationship between slope and carbon density on the Loess Plateau.

The slope of the Loess Plateau was positively correlated with the mean carbon density ($p = 0.01$), with correlation coefficients of 0.923 in 2000, 0.905 in 2005, and 0.907 in 2010. In addition, the carbon density exhibited three curve relations with slope, with a fitting degree of 0.963. The vegetation coverage in the area with a slope of 0~1° first decreased and then increased, and the total reduction was 887 km^2 from 2000 to 2010. The vegetation coverage in the area with a slope of 3~13° declined throughout the study period; from 2000–2005, it decreased by 210 km^2, and from 2005–2010, it decreased by 61 km^2. The vegetation coverage in the area with a slope of 14~26° remained stable over time. Furthermore, as the slope increased, the vegetation area decreased. Total carbon storage was negatively correlated with slope, with correlation coefficients of −0.718 in 2000, −0.719 in 2005, and −0.723 in 2010.

A change in altitude causes a change in climate factors, which impacts vegetation growth, development and physiological metabolism [24,25]. Accordingly, vegetation biomass and SOC were affected by altitude [21,26]. This study found that altitude in the Loess Plateau was significantly correlated with the average carbon density ($p < 0.01$), and the correlation coefficients were 0.747 in 2000, 0.543 in 2005, and 0.457 in 2010. However, total carbon storage was negatively correlated with altitude, with correlation coefficients of -0.481 in 2000, -0.487 in 2005, and -0.487 in 2010.

As altitude increased, land area decreased. Altitude and land area were significantly negatively correlated, with correlation coefficients of -0.498 in 2000, -0.498 in 2005, and -0.497 in 2010. Correspondingly, as altitude increased, forest area increased, and carbon density gradually increased (Figure 8). The Loess Plateau area is divided into seven grades according to altitude, <100 m, 100–200 m, 200–500 m, 500–1000 m, 1000–2000 m, 2000–3000 m and >3000 m, and the average carbon densities were 60.34, 59.79, 59.83, 64.92, 67.94, 67.85 and 73.32 t C/ha, respectively (Table 3).

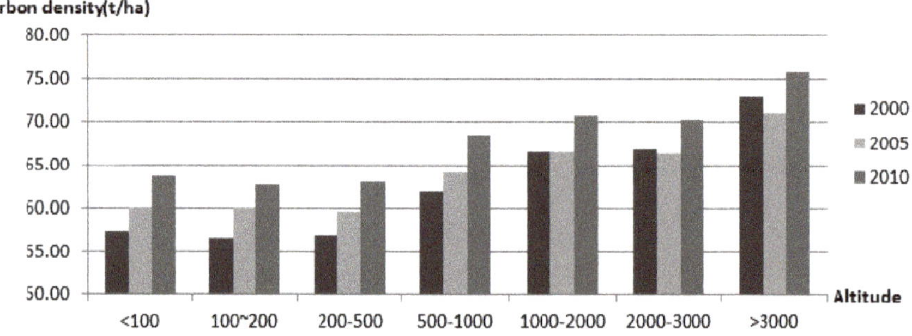

Figure 8. Relationship between altitude and carbon density.

Table 3. Carbon densities at different altitudes.

Altitude (m)	2000 (t/ha)	2005 (t/ha)	2010 (t/ha)	Average (t/ha)
<100	57.31	59.97	63.73	60.34
100~200	56.56	60.08	62.74	59.79
200–500	56.90	59.59	63.01	59.83
500–1000	62.02	64.17	68.57	64.92
1000–2000	66.53	66.61	70.68	67.94
2000–3000	66.94	66.41	70.21	67.85
>3000	72.94	71.11	75.92	73.32

4. Discussion

This study used raster data of global biomass and SOC along with sampling data to inversely calculate the aboveground biomass, belowground biomass, and SOC in the Loess Plateau. Then, the aboveground biomass, belowground biomass, SOC and humus carbon were determined. In addition, the accuracies of the inversion parameters were evaluated using the field-measured data, which indicated that the parameters were appropriate for further simulation and analysis. The InVEST model was used to simulate carbon storage, and the relationships between geographical factors and carbon storage, carbon density and land use/land cover type were analyzed. The accuracies of the results were verified using MYD17A2 and MOD17A2 products and soil and biomass survey data in the study area.

In this study, we found that the overall average value of carbon density ranged from 69.01 t/ha to 76.05 t/ha. The values are very different from those in other regions of China and global averages. This difference may be due to several factors. First, study differences in the number of data sets and sampling methods might be responsible. The Loess Plateau is a typical area and cannot reflect the world carbon distribution; its spatial heterogeneity might have contributed to the subtle differences between the woodlands of the study area and woodlands around the world. Second, the measurement accuracy of the NDVI data might have affected the uncertainties of the biomass estimates. Third, differences might be caused by natural selection expressed through plant morphology [27–30].

The spatial resolution of the simulated data in the Loess Plateau was 1 km, which is low, leading to uncertainty in the accuracy of the simulation result. The spatial resolution of the data used in the analysis of the geographical factors, such as elevation and slope, was 1 km, and there will be differences in the analyses of small terrain features or areas, which supports the view that differences in data sets and sampling methods may affect the results. Furthermore, the spatial resolution of the NDVI data and the processing methods affect the precision.

In addition, as reported in this study, not surprisingly, we found that carbon density or carbon storage was strongly related to geographical driving factors, thus supporting our hypothesis that carbon density is influenced by abiotic driving factors. Here, we explore and illustrate the relative contributions of geographical driving factors on carbon density.

Biomass is directly affected by plant traits and indirectly affected by soil properties [31,32]. In the present study, biomass was affected by slope and altitude because slope and elevation affect moisture and temperature, which affect plant growth. Geographical factors have direct effects on biomass owing to resource competition (e.g., light capture), and biomass is controlled more by land cover type and plant height, resulting in carbon storage [33,34]. In addition, soil properties have effects on biomass owing to the contribution of soil nutrients to the decomposition and mineralization of organic matter. In addition, SOC has strong effects on biomass through its effects on belowground biomass, which acts as a large nutrition source for root biomass storage [35–38].

There are a number of factors, such as light, temperature, rainfall, and plant species, that affect carbon sequestration. In this paper, the effects of only vegetation type, elevation and slope on carbon storage were evaluated. According to the carbon density analysis, the soil carbon density increased by 67.46 million tons of carbon, accounting for 25.94% of the total increase in carbon storage, over the ten-year study period. Soil carbon sequestration not only is a strategy to achieve food security through the improvement of soil quality but can also increase the global carbon reserves and improve the environment. Furthermore, soil carbon is a vitally important secondary product of increasing crop yield or enhancing the biomass of land cover. When carbon dioxide emissions into the air are reduced, soil carbon can improve and sustain biomass productivity. SOC is a very valuable natural resource. Regardless of climate change, soil carbon must be restored and improved. However, the close links between soil carbon sequestration and global food security or production and climate change cannot be overemphasized or ignored.

In addition, biomass increases have a major impact on carbon storage. Our results provide evidence to support the proposal in the Kyoto Protocol that carbon sequestration by afforestation or reforestation has the potential to partly offset carbon dioxide emissions from fossil fuel consumption, even though the increased carbon uptake is viewed as temporary. The Kyoto Protocol does not require commitments from developing countries, including China, but recent decreases in the rate of deforestation in China have already contributed to reduced carbon dioxide emissions. We believe that continuing the practice of nationwide afforestation and reforestation projects can contribute significantly to global carbon storage.

5. Conclusions

Carbon accumulation in the Loess Plateau region was simulated based on the InVEST model. The attribution analysis of the change of carbon storage identified the contributions of changes in carbon density and land cover to carbon storage. Moreover, the analyses of slope, elevation and other geographical factors revealed that these factors also affect carbon storage. This study found that the vegetation in the Loess Plateau gradually improved between 2000 and 2010. During this time, the biomass increased, and carbon storage slowly increased by 0.26 billion tons.

The overall change in land cover in the Loess Plateau was not large. The increases in the carbon densities of cultivated land and grassland were the most prominent. The carbon densities of areas with high carbon densities, such as woodland, shrubwood, forest and arable land, all increased. The carbon density of grassland decreased slightly, and the carbon densities of saline-alkali land and bog increased.

The study of carbon density and land cover change revealed that the contribution of land cover change to carbon storage was −1.0%. The effect of carbon density on carbon storage was 100.1%, which was mainly due to the implementation of the policy of returning farmland to forest (grassland) in the late 1990s [8]. This policy effectively promoted the restoration and reconstruction of vegetation in the Loess Plateau, including the existing vegetation area. In addition, it promoted the conservation and restoration of cropland and grassland and the restoration and reconstruction of low-vegetation-cover areas. These policies improved both the vegetation cover and the ecosystem carbon storage function. In addition, these factors are subject to climate change, which affects vegetation growth. Elevation and slope are also significantly correlated with carbon storage, which is directly related to meteorological and environmental factors, such as temperature, light, and water.

These findings reveal the driving factors influencing carbon storage and raise interesting questions: How is carbon storage regulated by vegetation growth and distribution in the Loess Plateau? Do plant biomass distribution, ecological environment and geographical factors drive carbon storage, and if so, what is the mechanism of interaction between these environmental driving factors and the storage of carbon? Future research will answer these questions.

Author Contributions: G.H. conceived and designed the experiments; Z.H. performed the experiments; Z.H. and G.H. analyzed the data; G.H. wrote the paper.

Acknowledgments: This research was supported in part by the National Natural Science Foundation of China under grants "Impact of Climate and Landuse Change on Water Quantity and Water Quality in the Mun River Basin (No. 41661144030)" and "Study on Water Resources Change and Adaptive Management in the Great Mekong River Basin (No. 41561144012)".

References

1. Jorgensen, L. *Newsletter of the Global Land Project International Project Office GLP News GLP Nodal Offices Newsletter of the Global Land Project International Project Office G Pn New E WS*; GLP: Taipei, Taiwan, 2010.
2. Foley, J.A.; DeFries, R.; Asner, G.P.; Barford, C.; Bonan, G.; Carpenter, S.R.; Chapin, F.S.; Coe, M.T.; Daily, G.C.; Gibbs, H.K.; et al. Global Consequences of Land Use. *Science* **2005**, *309*, 570–574. [CrossRef] [PubMed]
3. Rindfuss, R.R.; Walsh, S.J.; Turner, B.L.; Fox, J.; Mishra, V. Developing a science of land change: Challenges and methodological issues. *Proc. Natl. Acad. Sci. USA* **2004**, *101*, 13976–13981. [CrossRef] [PubMed]
4. Zhao, S.Q.; Liu, S.; Li, Z.; Sohl, T.L. Ignoring detailed fast-changing dynamics of land use overestimates regional terrestrial carbon sequestration. *Biogeosciences* **2009**, *6*, 1647–1654. [CrossRef]
5. Bennett, M.T. China's sloping land conversion program: Institutional innovation or business as usual? *Ecol. Econ.* **2008**, *65*, 699–711. [CrossRef]
6. Food and Agriculture Organization (FAO). *Global Forest Resources Assessment 2000 Main Report*; FAO Forestry Paper; FAO: Rome, Italy, 2010.
7. Breidenich, C.; Magraw, D.; Rowley, A.; Rubin, J.W. The Kyoto Protocol to the United Nations Framework Convention on Climate Change. *Am. J. Int. Law* **1998**, *92*, 315–331. [CrossRef]
8. Bohn, H.L. Estimate of Organic Carbon in World Soils. *Soil Sci. Soc. Am. J.* **1982**, *40*, 468–470. [CrossRef]
9. Eswaran, H.; Berg, E.V.D.; Reich, P. Organic Carbon in Soils of the World. *Soil Sci. Soc. Am. J.* **1993**, *90*, 269–273. [CrossRef]
10. Feng, J.; Hao, Y.; Qiguo, Z. Soil organic carbon and its influencing factors. *Soils* **2000**, *32*, 11–17.
11. Post, W.M.; Emanuel, W.R.; Zinke, P.J.; Stangenberger, A.G. Soil carbon pools and world life zones. *Nature* **1982**, *298*, 156–159. [CrossRef]

12. Qing, T.; Yong, X.; Yi, L. Spatial difference of land use change in Loess Plateau region. *J. Arid Land Resour. Environ.* **2010**, *24*, 15–21.

13. Ai, Z.; Chen, Y.; Cao, Y. Storage and allocation of carbon and nitrogen in *Robinia pseudoacacia* plantation at different ages in the loess hilly region, China. *Chin. J. Appl. Ecol.* **2014**, *25*, 333–341.

14. Li, J.; Wang, X.; Shao, M.; Zhao, Y.; Li, X. Simulation of biomass and soil desiccation of *Robinia pseudoacacia* forestlands on semi-arid and semi-humid regions of China's Loess Plateau. *Chin. J. Plant Ecol.* **2010**, *34*, 330–339.

15. Chen, X.; Ma, Q.; Kang, F.; Cao, W.; Zhang, G.; Chen, Z. Studies on the Biomass and Productivity of Typical Shrubs in Taiyue Mountain, Shanxi Province. *For. Res.* **2002**, *15*, 304–309.

16. Zhang, B.; Chen, C. Biomass and Production of Robinia Pseudoacacia Plantation in Hongxing Tree Farm of Changwu Country, Shaanxi Province. *Shaanxi For. Sci. Technol.* **1992**, *3*, 13–17.

17. Ruesch, A.; Gibbs, H.K. *New IPCC Tier-1 Global Biomass Carbon Map for the year 2000 [DB/OL]*; The Carbon Dioxide Information Analysis Center, Oak Ridge National Laboratory: Oak Ridge, TN, USA, 2008.

18. Carré, F.; Hiederer, R.; Blujdea, V.; Koeble, R. *Background Guide for the Calculation of Land Carbon Stocks in the Biofuels Sustainability Scheme Drawing on the 2006 IPCC Guidelines for National Greenhouse Gas Inventories*; Joint Research Center, European Commission: Luxembourg, 2010.

19. Paustian, K.; Ravindranath, N.H.; Amstel, A.V. *2006 IPCC Guidelines for National Greenhouse Gas Inventories*; Agriculture, Forestry and Other Land Use; IPCC: Geneva, Switzerland, 2006; Volume 4.

20. Sharp, R.; Tallis, H.; Ricketts, T.; Guerry, A.; Wood, S.; Chaplin-Kramer, R.; Nelson, E.; Ennaanay, D.; Wolny, S.; Olwero, N.; et al. *InVEST 3.1.2 User's Guide*; The Natural Capital Project: Stanford, CA, USA, 2015.

21. Li, J.; Ma, Y.L.; Luo, J.; Li, H.; Luo, Z.B. Nutrients and Biomass of Different-Aged Robinia Pseudoacacia Plantations in the Loess Hilly Region. Available online: http://www.cqvip.com/qk/97059x/201303/45949088.html (accessed on 14 May 2018).

22. Parton, W.J.; Schimel, D.S.; Cole, C.V.; Ojima, D.S. Analysis of Factors Controlling Soil Organic Matter Levels in Great Plains Grasslands1. *Soil Sci. Soc. Am. J.* **1987**, *51*, 1173–1179. [CrossRef]

23. Oren, R.; Ellsworth, D.S.; Johnsen, K.H.; Phillips, N.; Ewers, B.E.; Maier, C.; Schäfer, K.V.; McCarthy, H.; Hendrey, G.; McNulty, S.G.; et al. Soil fertility limits carbon sequestration by forest ecosystems in a CO_2-enriched atmosphere. *Nature* **2001**, *411*, 469–472. [CrossRef] [PubMed]

24. Chen, C.; Peng, H. Standing Crops and Productivity of the Major Forest-types at the Huoditang Forest Region of the Qingling Mountains. *J. Northwest For. Coll.* **1996**, *11*, 92–102.

25. Chen, Q.; Shen, C.; Yi, W.; Peng, S.; Li, Z. Progresses in soil carbon cycle researches. *Adv. Earth Sci.* **1998**, *13*, 555–563.

26. Huang, G.S.; Xia, C. MODIS-Based Estimation of Forest Biomass in Northeast China. *For. Resour. Manag.* **2005**, *4*, 40–44.

27. Mittelbach, G.G.; Steiner, C.F.; Scheiner, S.M.; Gross, K.L.; Reynolds, H.L.; Waide, R.B.; Willig, M.R.; Dodson, S.I.; Gough, L. What is the observed relationshio between species richness and productivity. *Ecology* **2001**, *82*, 2381–2396. [CrossRef]

28. Mccarthy, M.C.; Enquist, B.J. Consistency between an allometric approach and optimal partitioning theory in global patterns of plant biomass allocation. *Funct. Ecol.* **2007**, *21*, 713–720. [CrossRef]

29. Sun, J.; Cheng, G.W.; Li, W.P. Meta-analysis of relationships between environmental factors and aboveground biomass in the alpine grassland on the Tibetan Plateau. *Biogeosciences* **2013**, *10*, 1707–1715. [CrossRef]

30. Yan, L.; Zhou, G.; Zhang, F. Effects of different grazing intensities on grassland production in China: A meta-analysis. *PLoS ONE* **2013**, *8*, e81466. [CrossRef] [PubMed]

31. Xue, Z.; An, S.; Cheng, M.; Wang, W. Plant functional traits and soil microbial biomass in different vegetation zones on the Loess Plateau. *J. Plant Interact.* **2014**, *9*, 889–900. [CrossRef]

32. Yang, Y.; Dou, Y.; An, S. Environmental driving factors affecting plant biomass in natural grassland in the Loess Plateau, China. *Ecol. Indic.* **2017**, *82*, 250–259. [CrossRef]

33. Zuppinger-Dingley, D.; Schmid, B.; Petermann, J.S.; Yadav, V.; De Deyn, G.B.; Flynn, D.F. Selection for niche differentiation in plant communities increases biodiversity effects. *Nature* **2014**, *515*, 108–111. [CrossRef] [PubMed]

34. Kunstler, G.; Falster, D.; Coomes, D.A.; Hui, F.; Kooyman, R.M.; Laughlin, D.C.; Poorter, L.; Vanderwel, M.; Vieilledent, G.; Wright, S.J.; et al. Plant functional traits have globally consistent effects on competition. *Nature* **2016**, *529*, 204. [CrossRef] [PubMed]

35. Wang, Z.; Luo, T.; Li, R.; Tang, Y.; Du, M. Causes for the unimodal pattern of biomass and productivity in alpine grasslands along a large altitudinal gradient in semi-arid regions. *J. Veget. Sci.* **2013**, *24*, 189–201. [CrossRef]

36. Peña, M.A.; Duque, A. Patterns of stocks of aboveground tree biomass, dynamics, and their determinants in secondary Andean forests. *For. Ecol. Manag.* **2013**, *302*, 54–61. [CrossRef]

37. Goodness, J.; Andersson, E.; Anderson, P.M.; Elmqvist, T. Exploring the links between functional traits and cultural ecosystem services to enhance urban ecosystem management. *Ecol. Indic.* **2016**, *70*, 597–605. [CrossRef]

38. Li, N.; He, N.; Yu, G.; Wang, Q.; Sun, J. Leaf non-structural carbohydrates regulated by plant functional groups and climate: Evidences from a tropical to cold-temperate forest transect. *Ecol. Indic.* **2016**, *62*, 22–31. [CrossRef]

Article

Instability Analysis of Supercritical CO_2 during Transportation and Injection in Carbon Capture and Storage Systems

Il Hong Min [1], Seong-Gil Kang [2] and Cheol Huh [1,*]

[1] Department of Convergence Study on the Ocean Science and Technology, Ocean Science and Technology School, Korea Maritime and Ocean University, Busan 49112, Korea; ihmin@kmou.ac.kr
[2] Korea Research Institute of Ships and Ocean Engineering, Daejeon 34103, Korea; kangsg@kriso.re.kr
* Correspondence: cheolhuh@kmou.ac.kr; Tel.: +82-51-410-5247

Received: 16 July 2018; Accepted: 3 August 2018; Published: 6 August 2018

Abstract: Captured CO_2 is in a subcritical state, whereas CO_2 deep underground is in a supercritical state because of the high geothermal heat and pressure. The properties of CO_2 can change rapidly at the critical point and in the near-critical region during the transportation and injection process. This study aims to identify the instabilities in the CO_2 flow in these regions, along with the causes and effects, during the transportation and injection process, and propose relevant design specifications. Thus, the critical points and near-critical region of CO_2 flow were numerically analyzed. The unstable region is presented in terms of temperature and pressure ranges, and the changes in the CO_2 properties in this region were analyzed. In the unstable region, the sudden change in density was similar to the density wave oscillation of a two-phase flow. The CO_2 stability map we obtained and the stability map of supercritical water show similar trends. Flow instability was also found to occur in standard CO_2 transportation pipelines. We demonstrate that flow instability in CO_2 transportation and injection systems can be avoided by maintaining the proposed conditions.

Keywords: supercritical CO_2; flow instability; stability map; CO_2 pipeline; carbon capture and storage (CCS)

---

1. Introduction

Carbon capture and storage (CCS) is a technology used to capture CO_2 emissions from power plants or industrial facilities to reduce greenhouse gas emissions; the gas is subsequently injected and stored in depleted oil and gas reservoirs or saline aquifers [1].

The total value chain of CCS technology comprises capture, transportation, injection, and storage. The CO_2 captured in power plants or industrial facilities is in a subcritical state. However, the CO_2 stored deep underground is maintained in a supercritical state because of the high geothermal heat and pressure in the reservoir. This implies that the operating conditions associated with the critical point and near-critical regions may occur during the transportation and injection processes, which are intermediate steps, because of the phase difference between the captured CO_2 and stored CO_2. In a previous study [2–4] by the authors of this paper, a depleted gas field in the East Sea of Korea was studied as a storage reservoir. Near-critical point flow and supercritical state flow of CO_2 were found to occur in the topside pipeline of the offshore platform and the upper part of the injection wellbore at a specific injection point, thereby causing flow instability in the pipeline [5]. Such flow instability can cause an additional pressure drop in the pipeline or disable flow control. Therefore, the near-critical point and supercritical state flows of CO_2 should be considered in the initial design stage to allow the safe and economic injection and storage of CO_2 in CCS systems. To this end, the properties and flow characteristics of CO_2 at the critical point and in the near-critical region must be studied thoroughly.

The purpose of this study is to identify the instability in the CO_2 flow at the critical point and in the near-critical region during the transportation and injection processes for CCS, to analyze the cause and effect of the flow instability, and to propose design specifications that reflect the results of the analysis.

Previous studies on the properties and flow instability of supercritical fluids similar to the case employed in this study can be categorized into four areas. The first area of study pertains to the critical and pseudocritical states of single-component fluids. Pioro and Mokry [6] presented the terms and definitions for critical and supercritical fluids. The saturation line of fluids such as pure water or CO_2 is a single line, and the pressure and temperature of CO_2 at the critical point are 73.8 bar and 30.9 °C, respectively. In a particular part of the supercritical region above the critical point, the properties of the single-component fluid change abruptly at a particular pressure and temperature, and this is defined as the pseudocritical point. Furthermore, more than one pseudocritical point can exist, and the line connecting these points is called the pseudocritical line [7]. Every pseudocritical temperature in the pseudocritical line has a corresponding pseudocritical pressure. Figure 1 shows the phase envelope of CO_2 which was calculated using NIST's REFPROP [8]. As shown in Figure 1, the pseudocritical line tends to extend from the saturation line of CO_2 in the subcritical state to the supercritical state via the critical point. The specific heat, density, and enthalpy of the supercritical CO_2 change drastically near the pseudocritical line [9]. The abrupt behavior of the properties at the pseudocritical line tends to decrease further from the critical point [10]. In the case of the phase envelope of natural gas, as shown in Figure 2 (which was calculated using NIST's REFPROP [8]), the saturation line takes the form of a parabola because the gas comprises a hydrocarbon mixture, unlike CO_2.

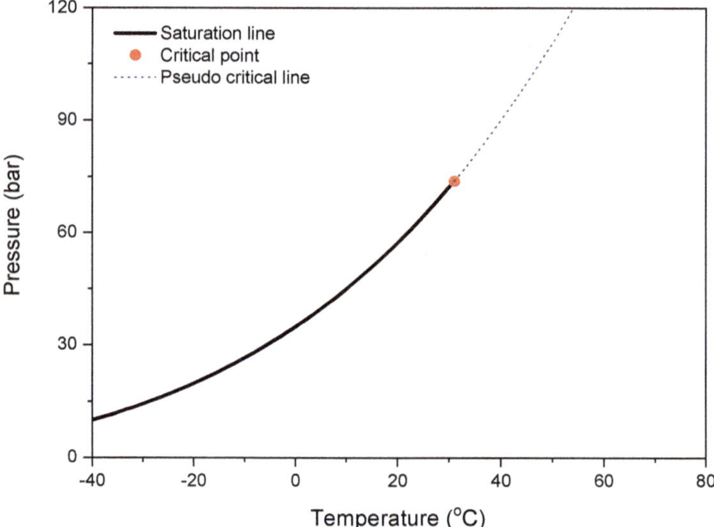

Figure 1. Phase envelope of pure CO_2.

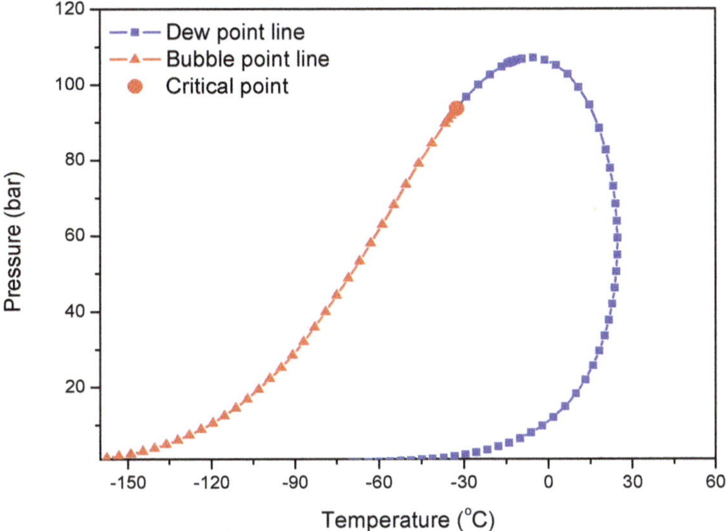

Figure 2. Phase envelope of natural gas.

Although the phase behavior of natural gas has been thoroughly studied, studies of the phase behavior of CO_2 are lacking. Shoaib et al. [11] investigated the effects of the operating conditions on the dew point and condensation rate of natural gas and developed two quadratic equations to relate them to the operational variables of the dew point control unit (DPCU). Guerrero-Zárate et al. [12] developed and validated a new algorithm to determine the critical point of a natural gas mixture. Wang et al. [13] derived implicit curve-fitting equations to increase the speed of calculation of natural gas properties in the supercritical pressure region. In summary, there is an abundance of knowledge and experience regarding the transportation of natural gas through pipelines [14–17]. However, it is difficult to analyze the phase transition in the transport of CO_2, which is the objective of this study, because of the single saturation line. Moreover, the behavior of near-critical point flow is significantly more complicated than that of a natural gas flow.

The second area of study relates to the flow of supercritical CO_2 and the application of supercritical and pseudocritical CO_2 to thermodynamic cycles. Ma et al. [18] conducted a study to improve the performance of the Brayton cycle using the thermodynamic properties of pseudocritical CO_2. Shao and Zhang [19] considered the saturation temperature of subcritical state CO_2 as the pseudocritical temperature of supercritical state CO_2 and proposed the possibility of using supercritical state CO_2 in the CO_2 cycle by extending the vapor pressure equation. Gupta et al. [20] established a correlation between the heat transfer characteristics based on the experimental pseudocritical CO_2 data obtained for a supercritical region flow in a vertical tube. They applied the correlation to the Brayton gas turbine cycle. In addition, active studies have been carried out to apply the rapidly increasing specific heat and heat transfer coefficient of pseudocritical CO_2 in the supercritical region to the working fluid of the cycle [21–24]. However, most studies have focused solely on the analysis of thermodynamic properties, with relatively few studies considering the flow instability of supercritical state CO_2. Furthermore, because the experimental cycle used for analysis is small compared to a CCS transportation system, wherein the pipelines are on the order of several tens to several hundreds of kilometers, the results are limited in their extension to the CO_2 transportation pipelines employed in CCS technologies.

The third area of study is the flow instability of fluids at supercritical and near-critical points in the natural circulation loops used to cool nuclear power systems. Archana et al. [25] experimented with and analyzed the transient flow of supercritical CO_2 in a natural circulation loop. Chen et al. [26]

performed a numerical analysis on the flow instability arising from rapid changes in the properties of pseudocritical state CO_2 in a natural circulation loop. Other experiments and numerical analyses have been conducted to analyze the flow instability of supercritical CO_2 in a natural circulation loop [27–29]. These studies have investigated the instability of supercritical CO_2 flow, similar to that in this study. However, for the natural circulation loop considered in their studies, the driving force of the flow is the buoyancy arising from the density difference, whereas the driving force of the flow in a long-distance CO_2 pipeline is the pressure difference. Therefore, it is difficult to use the results of their studies for the CO_2 transportation and injection processes in CCS. Moreover, the scale of the system is different.

The fourth area of study regarding the flow instability of supercritical fluids is the flow instability of supercritical water caused by the density wave oscillations that occur near the pseudocritical line. The density and compressibility factor of supercritical water change suddenly along the pseudocritical line, resulting in unstable fluid behavior [30]. Ambrosini [31] argued that the flow instability of supercritical water in heated channels is similar to the flow instability caused by the two-phase flow of subcritical water. The similarity in the obtained stability maps validated the results. Xiong et al. [32] investigated the flow instability of supercritical water flowing in a vertical pipeline and analyzed the main parameters causing flow instability. However, the critical pressure and temperature of water differ from those of CO_2, and the changes in the properties are different in terms of the critical temperature and pressure. Therefore, it is difficult to apply these values to the CO_2 transportation and injection processes of a CCS system.

As mentioned above, to implement CCS technology efficiently, the near-critical state flow and the supercritical state flow of CO_2 must be considered. However, existing studies have not thoroughly analyzed the flow instability caused by near-critical or supercritical CO_2 flows. Therefore, in this study, a numerical analysis was conducted, focusing on the flow instability caused by the operating conditions associated with near-critical point and supercritical states in a CO_2 transportation and injection system.

2. Method of Numerical Analysis

2.1. CCS Chain Configuration

A CO_2 transportation and injection system model was made using four components: a subsea pipeline, a riser, a topside pipeline, and a wellbore. It was assumed that the CO_2 captured in onshore power plants is temporarily stored in a hub terminal near the storage site [3,33]. The CO_2 at the hub terminal is then transported through the subsea pipeline to the sea of the reservoir, and the transported CO_2 is elevated to the offshore platform using the riser. The injection equipment on the topside of the offshore platform comprises a heater, a choke valve, metering devices, monitoring devices, and isolation devices. The CO_2 is transferred from the injection facility on the offshore platform to the sub-seabed through the injection riser and is injected from the sub-seabed to the reservoir using the injection wellbore. Figure 3 shows the overall configuration of the transportation and injection system [3,33].

In this study, the possibility, causes, and effects of flow instability when CO_2 in the topside pipeline and injection wellbore flows in near-critical and supercritical states were analyzed using numerical analysis. OLGA 2014.1 [34] was used for this purpose. The CO_2 in a CCS system is assumed to have a purity of 99% or more. The properties of pure CO_2 were used to determine the behavior of supercritical state CO_2 near the critical point from a single saturation line accurately. The single component module (OLGA) was used to simulate the behavior of pure CO_2 in the system [35].

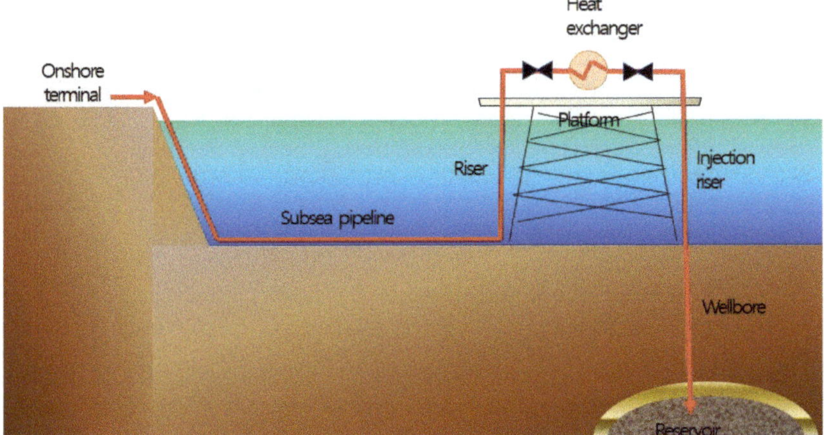

Figure 3. Schematic of the offshore CO_2 transportation and injection system.

The numerical methods and mathematical descriptions were described in the previous study [35] of the authors. The pipeline was designed to be installed in the sub-seabed approximately 60 km from the coastal terminal to the sea where the platform is installed. The riser was designed to have a length of 155 m, measuring from the end of the seabed pipeline to the topside of the offshore platform and considering the seawater depth of the reservoir area. The topside pipeline was approximately 1 km in length and was equipped with isolation valves, a heat exchanger for heating, and choke valves. The injection wellbore comprises a seawater injection riser, which connects the platform to the sub-seabed, and an underground injection wellbore. The length of the underground injection wellbore was designed such that it could be installed vertically to a depth of approximately 2438 m, measuring from the sub-seabed to the storage reservoir.

2.2. Calculation Conditions

Table 1 lists the main calculation conditions for the transportation and injection system proposed in this study. The inside diameter of the pipeline was 8 in (20.32 cm). The same value was used for the entire system including for the subsea pipeline, riser, topside pipeline, and injection wellbore. The inner diameter was selected considering the flow rate, pressure drop, heat transfer, and erosional velocity ratio (EVR), as given in the previous study of the authors [33].

Table 1. Calculation conditions for the numerical simulation.

Calculation Condition	Design Value
CO_2 composition	Pure CO_2
CO_2 flow rate	31.5 kg/s
Pipeline diameter	8 in
Topside arrival temperature of CO_2	5.7 °C
Reservoir temperature	97.8 °C
Length of the topside pipeline	1000 m
Length of wellbore	2438 m

For the numerical analysis of the CO_2 flow in the transportation and injection system, the effect of the increase in the reservoir pressure caused by CO_2 injection on the system should be considered. The pressure of the reservoir increases over the injection period because of the accumulation of CO_2 in the reservoir. The correlation between the pressure change in the reservoir and the bottom hole and the injection rate of CO_2 over time was obtained through reservoir simulation [36].

As shown in Figure 4, the bottom hole pressure increased over an injection period of 120 months because of the accumulation of injected CO_2. The pressure of the entire transportation and injection system increased because of the increased bottom hole pressure, eventually making the flow in a specific section of the system enter the supercritical state.

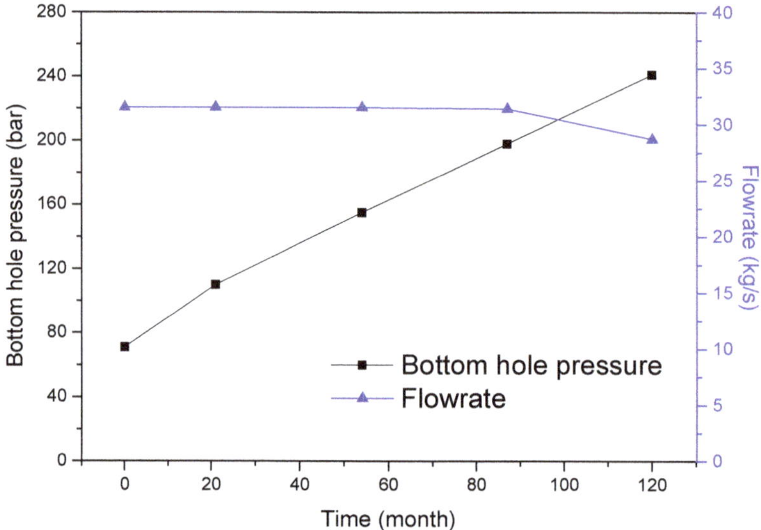

Figure 4. Variations in the reservoir pressure, bottom hole pressure, and flow rate over the injection period.

In the previous study by the authors, the operating pressure near the CO_2 critical pressure (Pc = 73.8 bar) occurred at the topside during the injection period (approximately 54 months), and the temperature of the CO_2 arriving at the topside was calculated to be 5.7 °C [5]. When the topside CO_2 was heated to a near-critical temperature (Tc = 30.9 °C) using the heater, the flows in the topside pipeline and injection wellbore became unstable.

To analyze the CO_2 flow instability, simulations were performed by varying the temperature and pressure. The bottom hole pressure was controlled to allow the operating conditions to reach the critical pressure in the topside pipeline. The heating temperature of the topside heater at each bottom hole pressure was set near the critical temperature (30 to 40 °C). In particular, numerical analysis was carried out for 20 cases, wherein the heating temperature of the heater was increased by 1 °C from 30 to 40 °C in each of the 10 cases and was decreased by 1 °C from 40 to 30 °C in an additional 10 cases. The generation of the transient state was checked while maintaining the temperature condition for 100 h, i.e., by calculation for a sufficient time. Furthermore, to eliminate the influence of the rapid temperature change, the time required to increase or decrease the temperature by 1 °C was set to 10 h.

Figure 5 shows the heating temperature of the heater with respect to the simulation time. Because detailed information concerning the internal configuration of the heater is unavailable at this stage, it was assumed that the pressure drop generated through the heater was negligible. In addition, the pressure, temperature, and density of the CO_2 at the heater outlet were analyzed to verify the flow instability after heating using the topside heater. Table 2 lists the bottom hole pressure, topside operating pressure, and heater set temperature for each numerical analysis case.

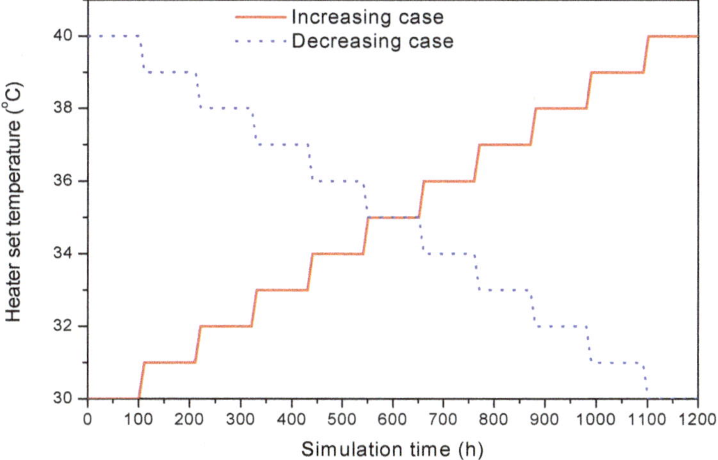

Figure 5. Variation in the heater set temperature with respect to the simulation time.

Table 2. Operating conditions for each calculation case.

Case	Bottom Hole Pressure (bar)	Topside Operating Pressure (bar)	Increase Heater Set Temperature (°C)	Decrease Heater Set Temperature (°C)
1	170	70–76		
2	180	72–78		
3	190	73–79		
4	200	74–80		
5	210	77–83	30–40	40–30
6	220	78–84		
7	230	80–86		
8	240	82–88		
9	250	84–90		
10	260	85–91		

3. Results

3.1. Identification of the Unstable Region

Figure 6 shows the calculation results for case 4 as an example. The density, pressure, and temperature of CO_2 were analyzed at the outlet of the heater. The outlet conditions of the heater include a pressure of 79 bar and a temperature of 37 °C, which are higher than those at the critical point of CO_2 (Pc = 73.8 bar and Tc = 30.9 °C). Under these operating conditions, the heater set temperature gradually decreases and approaches the critical point operating condition. When the simulation time is longer than 250 h, the operating pressure and temperature of the heater outlet are approximately 75 bar and 35 °C, respectively. At this point, the density and pressure of the CO_2 at the heater outlet start to vary. In this study, the operating conditions associated with the flow instability at the near-critical point were confirmed for all cases listed in Table 2. In every case, the flows destabilized because of the fluctuations in the density, pressure, and temperature when the operating conditions were closer to the critical point. In this study, the conditions for the occurrence of flow instability were determined when the density variation was ± 50 kg/m^3 or more at a steady state. Furthermore, the pressure and temperature at this time were considered operating conditions, causing flow instability in the pipeline. The temperature and pressure at which the flow started to destabilize were obtained for each case listed in Table 2. The region of flow instability was determined for the critical point, near-critical region, near the pseudocritical line region, and supercritical region.

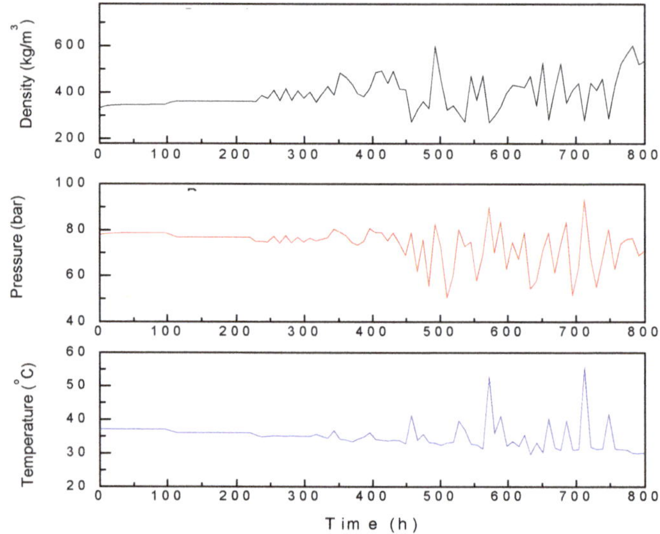

Figure 6. Simulation results of case 4 (reduction in heater temperature).

The pressure and temperature at which the flow begins to destabilize can be found using the pressure–temperature diagram of CO_2 by combining the aforementioned simulation results. Figure 7 shows the results, wherein the pressure and temperature ranges are indicated. The unstable region is highlighted as the colored region in Figure 7. Furthermore, the instability boundaries which were obtained through the numerical simulations are also specified.

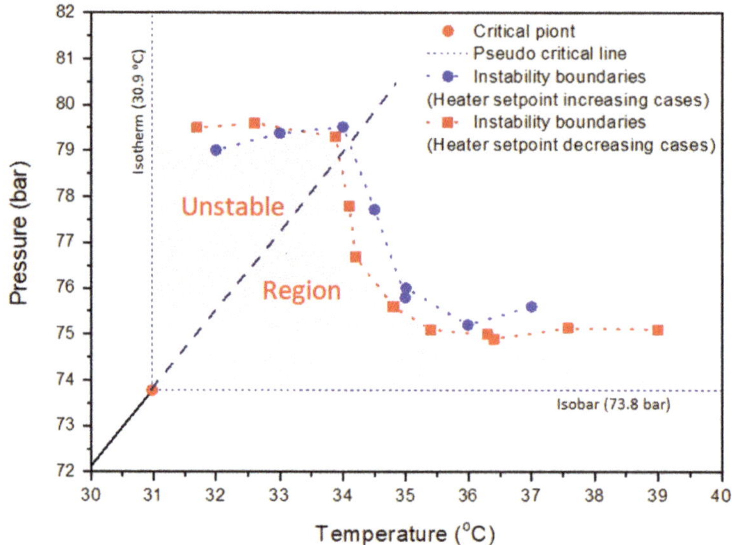

Figure 7. Unstable region of supercritical CO_2.

The unstable region appears to be similar in both heating temperature cases, i.e., increasing and decreasing, and is near the critical point of CO_2. In this study, this region is defined as an unstable region. Among the regions shown in Figure 7, flow instability was not observed in the topside and injection wellbore when the pressure and temperature are above those at the boundary of the regions.

Such an unstable region exhibits several characteristics. For example, the corresponding pseudocritical pressure is approximately 79.8 bar at a pseudocritical temperature of approximately 34.5 °C. The unstable regions can be distinguished using this value. In other words, the pressure should be higher than 79.8 bar ($P > Ppc = 79.8$ bar) to prevent instability at a temperature lower than the pseudocritical temperature of approximately 34.5 °C ($T < Tpc = 34.5$ °C). In addition, the pressure should be higher than 77 bar ($P > Ppc = 77$ bar) to avoid instability above the pseudocritical temperature of approximately 34.5 °C ($T > Tpc = 34.5$ °C).

Thus, the existence of an unstable CO_2 flow region in the pipeline suggests that the unstable flow must be considered in the design and operation of the entire CCS value chain in which CO_2 flows in the supercritical state. Considering the unstable flow, a specific range of operating conditions must be defined for system design and operation. Moreover, analytical results for critical and supercritical CO_2 that correspond to the unstable flow are required. The results of this study identify the specific region near the critical point where the CO_2 flow is unstable. CO_2 near the critical point is known to exhibit sudden changes in its properties even on small pressure and temperature variations based on existing studies and experiments. However, if the CO_2 transportation and injection system is operated in an unstable region, the cause of flow instability should be analyzed in detail. Thus, the changes in the properties of CO_2 in the unstable region and near the pseudocritical line were analyzed, as explained in detail in the next section.

3.2. Effect of Property Changes Near the Pseudocritical Region

Figures 8 and 9 show the unstable regions in the temperature–density and the pressure–density diagrams of CO_2, respectively. The unstable region is highlighted as the colored region. Furthermore, the instability boundaries which were obtained through the numerical simulations are also specified. The density of CO_2 changes rapidly even with a slight temperature change at the critical point and in the near-critical region, as shown in Figure 8.

Similarly, as shown in Figure 9, a rapid density change was observed, even with a slight pressure change at the critical point and in the near-critical region. However, the density change behavior occurs not only at the critical point and in the near-critical region but also in the supercritical region. For example, as shown in Figure 8, the density of CO_2 changes rapidly if there is a slight change in the pressure or temperature at supercritical conditions, i.e., a pressure of 78 bar and a temperature of 33.5 °C. Assuming that the temperature changes from 33 to 34 °C at a constant pressure of 78 bar, the change in the density of CO_2 is more than 300 kg/m^3. The unstable region includes an area with significant density variations because of temperature or pressure changes. The sudden changes in the properties are evident along the pseudocritical line, as shown in Figure 7 and reported in previous studies. In particular, the properties change abruptly near the pseudocritical line in the unstable region. In other words, in the CO_2 transportation and injection system analyzed in this study, if the operating pressure of the topside is in the unstable region and the heater outlet temperature is set to a temperature within the unstable region, the pressure or temperature of the heated CO_2 may change slightly because of the pressure drop or heat transfer in the pipeline, resulting in a significant change in the density.

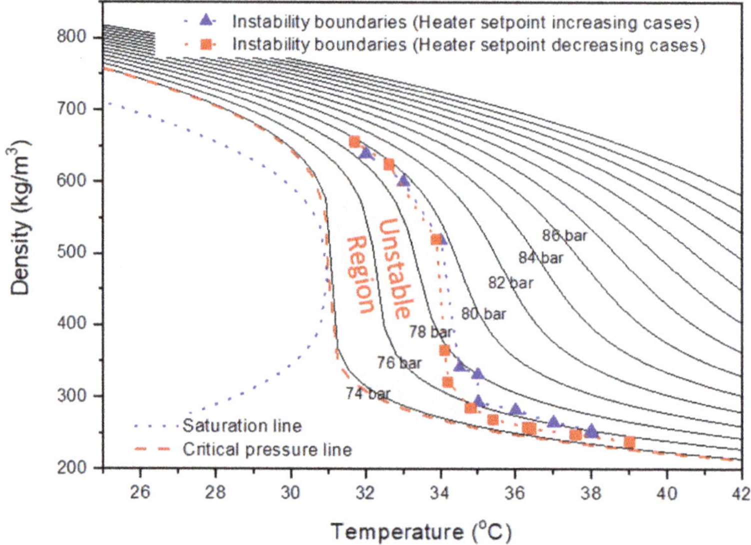

Figure 8. Temperature–density diagram of CO_2 with an unstable region.

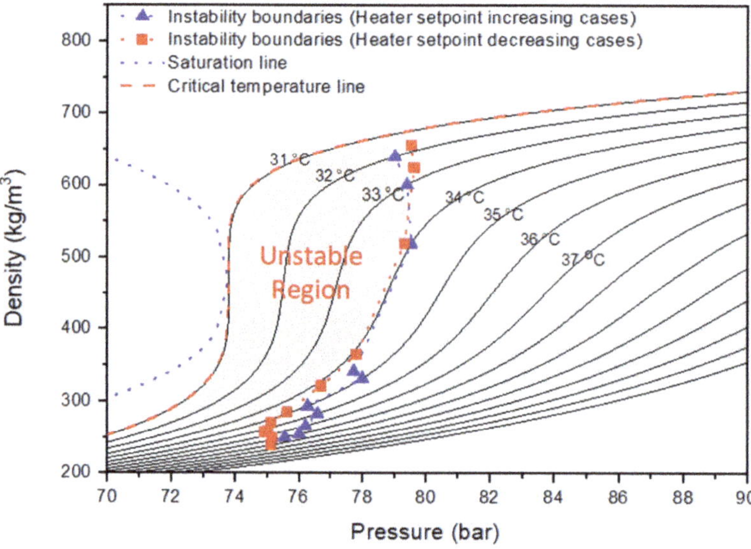

Figure 9. Pressure–density diagram of CO_2 with an unstable region.

To analyze the density behavior of CO_2 in the unstable region in more detail, the density gradient of CO_2 was plotted with respect to the pressure and temperature, as shown in Figures 10 and 11. The pressure and temperature ranges, shown in Figures 10 and 11, are within the previously obtained unstable region. Along the pseudocritical line in the unstable region, the density of CO_2 changes significantly depending on the pressure and temperature. The density gradient is considerable at the critical point, whereas it decreases as the temperature and pressure move away from the critical point, i.e., outside the unstable region. In the aforementioned region, shown in Figure 7, the operating conditions, i.e., a pressure of 80.7 bar and a temperature of 35 °C, are located on the pseudocritical line; however, they are outside the unstable region. If the same logic is applied to the relationship shown in Figures 10 and 11, there is no notable density change in the pseudocritical line above 79.8 bar and 34.5 °C.

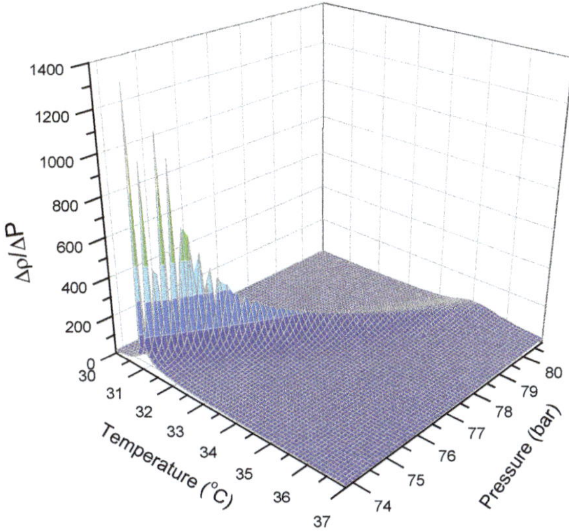

Figure 10. Density gradient of CO_2 with respect to pressure.

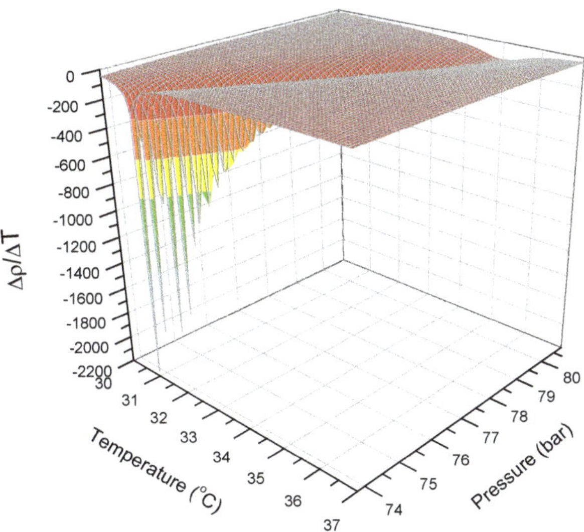

Figure 11. Density gradient of CO_2 with respect to temperature.

As shown in the CO_2 density gradient with respect to the pressure in Figure 10, the pressure and temperature at which the density gradient becomes a maximum lie on the pseudocritical line shown in Figure 7. At the critical point and in the near-critical region of CO_2, the density gradient arising from the pressure increases sharply up to approximately 1000 kg/m^3·bar. On the other hand, the density gradient of CO_2 with respect to the temperature, as shown in Figure 11, decreases sharply to approximately 2100 kg/m^3·°C with increasing temperature. Figures 10 and 11 show that the density gradient of CO_2 on the pseudocritical line is more susceptible to temperature than to pressure.

The above results show that the density gradient of CO_2 with respect to the changes in the pressure and temperature changes rapidly in the unstable region. If this behavior occurs in the CO_2 transportation and injection system, a phenomenon similar to density wave oscillations arising from the difference in the densities of the liquid and vapor phases in two-phase flow may occur, even though CO_2 is in a single-phase supercritical state. This phenomenon is analyzed in Section 3.3.

3.3. Analysis of the Instability

Based on the results in the preceding section, CO_2 in the supercritical state near the critical point and in the unstable region shows an unstable state with rapid changes in its properties, even when the temperature and pressure change slightly. If these rapid changes in properties occur in the CO_2 transportation pipeline and injection wellbore, they can cause severe flow instability, possibly resulting in loss of flow control. In this section, the cause of the above-mentioned flow instability in the unstable region is analyzed.

Generally, in a two-phase flow system, density wave oscillations are due to the difference in the densities between the high-density liquid phase and the low-density vapor phase, and this can cause a sudden pressure drop or perturbation in the flow velocity, resulting in flow instability in the system [37]. However, the fluid density along the pseudocritical line can change rapidly even in a supercritical fluid, and this can also cause a phenomenon similar to density wave oscillation. Studies have shown that flow instability arising from the density difference occurs even in a supercritical single-phase water flow, which is similar to density wave oscillations in the two-phase flow of subcritical water [31]. The CO_2 flow instability in the supercritical state and in the near-critical region occurs in the single-phase supercritical region and is, thus, different from the density wave oscillation of the existing two-phase system. However, based on the results reported in the previous chapter, the density change of CO_2 occurring at the pseudocritical line in the unstable region appears to be similar to the density change behavior occurring at the saturation line. This resulted in a behavior similar to the flow instability arising from the phase change in a two-phase flow system.

In the supercritical CO_2 single-phase flow of this study, a phenomenon similar to the flow instability that occurs in two-phase flow because of the phase change of CO_2 was observed. Therefore, the calculation results of this study were applied to a stability map using two non-dimensional numbers: the sub-pseudo-critical number, N_{SUBPC}, and the trans-pseudo-critical number, N_{TPC}. The two dimensionless numbers have been used to describe the instability boundaries [38,39]. These two dimensionless numbers have been modified such that the phase change number, N_{PCH}, and the subcooling number, N_{SUB}, of the existing subcritical fluid stability map could be applied to a supercritical fluid [38]. A stability map that distinguishes between the stable and unstable flow regions can be obtained by calculating and plotting N_{SUBPC} (sub-pseudo-critical number) and N_{TPC} (trans-pseudo-critical number) of the supercritical fluid. The exact formulae are given in Equations (1) and (2), respectively.

$$N_{\text{SUBPC}} = \frac{\beta_{pc}}{C_{p,pc}}\left(h_{pc} - h_{in}\right) \tag{1}$$

$$N_{TPC} = \frac{\dot{Q}}{\dot{m}}\frac{\beta_{pc}}{C_{p,pc}} \tag{2}$$

Here, β_{pc} is the coefficient of volume expansion at the pseudo-critical point, $C_{p,pc}$ is the specific heat at the pseudocritical point, h_{pc} is the enthalpy at the pseudocritical point, h_{in} is the enthalpy of CO_2 at the topside heater inlet, $\dot{Q}$ is the heat flow rate input from the topside heater, and $\dot{m}$ is the mass flow rate of CO_2 in the system.

In this study, the same temperature of the heater inlet condition, 5.7 °C, was applied, which is the topside arrival temperature, and h_{in} was calculated using the operating pressure and inlet temperature of every calculation case. The CO_2 properties were calculated using NIST's REFPROP [8]. For the flow rate of CO_2 used, a constant value of 31.5 kg/s was applied to the system. Furthermore, the amount of heat input to the heater was calculated for each simulation case.

The two non-dimensional numbers applied to this system were calculated separately and categorized into three groups: the operating conditions of the unstable region boundaries obtained above, operating conditions of the unstable region wherein the flow is unstable, and operating conditions outside the unstable region where the flow is stable.

The N_{SUBPC} value increases with increasing difference between the temperature corresponding to the pseudocritical enthalpy of the heater operating pressure and the temperature at the heater inlet. The N_{TPC} value tends to increase as the heater increasingly heats the CO_2. Figure 12 shows that the trend in the flow instability of supercritical CO_2 is clearly different at the boundary of the previously obtained unstable region.

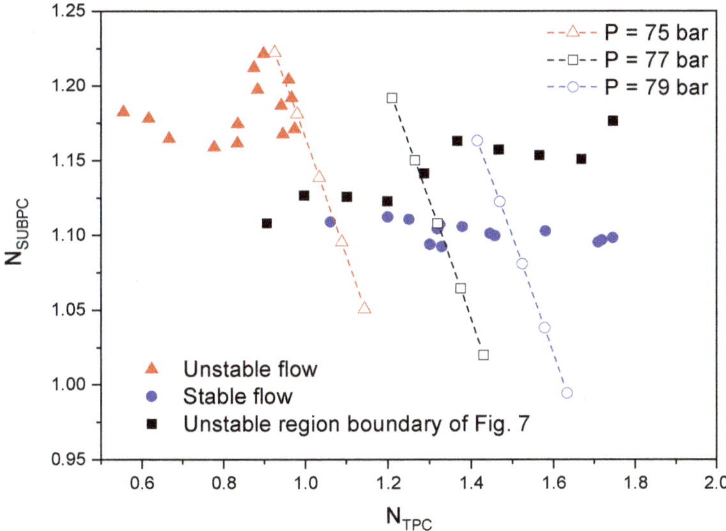

Figure 12. Stability map of supercritical CO_2.

As shown in Figure 12, the stability map obtained using the two dimensionless numbers indicates that the flow in the system becomes more unstable with increasing subcooling at the heater inlet, and the flow in the system becomes more stable as more heat is input to the heater. The effect of the inlet subcooling and the supplied heat at constant pressure is investigated. Table 3 shows the calculation conditions. As the subcooling increases (N_{SUBPC} increase), the flow becomes unstable. Furthermore, as more heat is supplied (N_{TPC} increase), the flow becomes more stable.

When the results, shown in Figure 12, are compared to the stability map of supercritical water obtained by Ambrosini and Sharabi [39], differences in the ranges of N_{SUBPC} and N_{TPC} values are observed; however, the flow instability trend was similar. The difference between the ranges of the two dimensionless numbers was caused by the significant differences in the properties of the supercritical water and supercritical CO_2 depending on the pressure and temperature.

Table 3. Simulation conditions for constant pressure case.

Inlet Pressure (bar)	Inlet Temperature (°C)	Mass Flowrate (kg/s)	Supplied Heat (kW)
75	5	31.5	3400
	7		3600
	9		3800
	11		4000
	13		4200
77	5	31.5	4400
	7		4600
	9		4800
	11		5000
	13		5200
77	5	31.5	5200
	7		5400
	9		5600
	11		5800
	13		6000

3.4. Case Study on CO_2 Pipeline Transport

A case study was conducted on a standard CO_2 pipeline to determine whether the results of this study were specific to a system that is heated by a topside heater, as in this study. To this end, a numerical analysis was carried out by modeling a 10 km-long horizontally installed pipeline used to transport CO_2. Table 4 outlines the calculation conditions for the numerical analysis of the CO_2 pipeline in each case. The same values for the pipeline diameter and flow rate as those of the existing system were used. The initial pressure and temperature in the pipeline were all outside the unstable region of the pipeline. As the simulation time passed from 10 h to 40 h, the outlet pressure of the pipeline decreased, and the operating conditions were within the unstable region. Through this process, we analyzed whether the flow instability occurs in the pipeline when the pipeline inlet with a high pressure operating condition is located outside the unstable region and when the pipeline outlet with a low pressure operating condition is located inside the unstable region.

Table 4. Simulation conditions for each horizontal pipeline case.

Case	Time (h)	Pipeline Length (km)	Temperature (Inlet–Outlet) (°C)	Pressure (Inlet–Outlet) (bar)
1	10	10	34.0–32.6	88.5–83.0
	20			86.7–81.0
	30			85.0–79.0
	40			81.0–77.1
2	10	10	36.0–34.0	88.2–82.0
	20			86.6–80.0
	30			85.0–78.0
	40			81.3–76.0
3	10	10	39.0–35.9	89.2–82.0
	20			87.8–80.0
	30			85.3–77.0
	40			84.9–75.0

Figure 13 shows the pressure–temperature diagram of CO_2, showing the pressure and temperature calculated at intervals of 10 h in the pipeline for each case study. In all three cases, the temperature and pressure fluctuated because the operating conditions near the pipeline outlet are located in the unstable region. This fluctuation shows that the flow instability spreads to all pipeline sections. Consequently, the abnormal pressure and temperature profiles were shown in 40 h, see Figure 13. Concerning the cause of the flow instability mentioned above, the rapid changes in the CO_2 properties at the pipeline outlet affected the entire pipeline. This result suggests that the unstable CO_2 region at the critical point and in the near-critical region obtained from this study affects the flow instability of general pipelines. In other words, the unstable behavior of CO_2 in the unstable region not only occurs when it is heated by a heat exchanger, such as the heater used in this study, but also in more general cases.

If the temperature and pressure of the fluid are within the unstable region in a specific section of the CO_2 transportation and injection system, the conditions are sufficient to cause flow instability in the entire system. Therefore, the operating conditions of the CO_2 transportation and injection system should be designed in such a manner to avoid the unstable region, as shown in the results of this study.

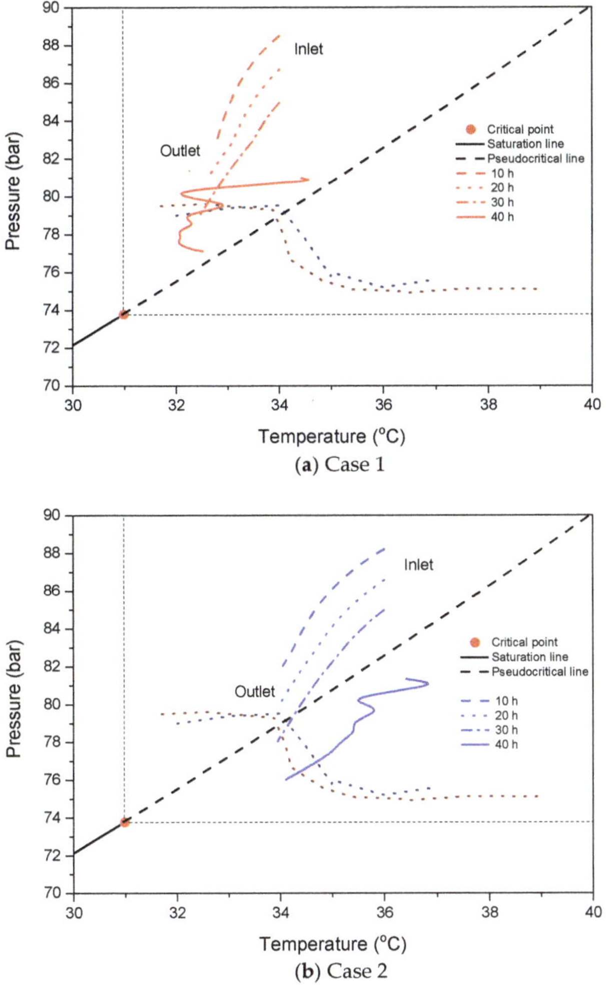

(**a**) Case 1

(**b**) Case 2

Figure 13. *Cont.*

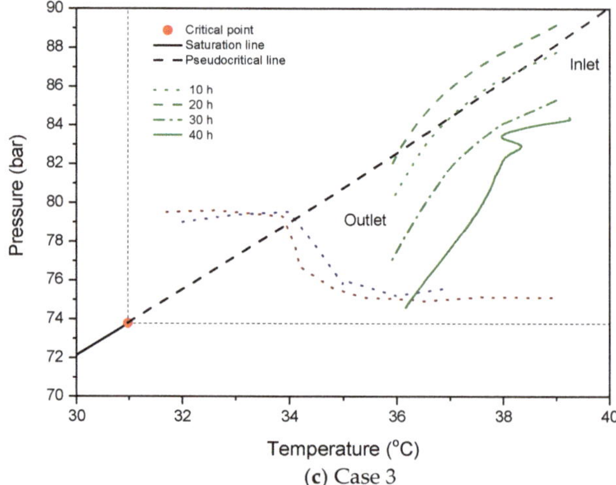

Figure 13. Pressure–temperature profile of the horizontal pipeline.

4. Conclusions

The CO_2 captured in CCS systems is in a subcritical state at room temperature and atmospheric pressure and is stored in a supercritical state in the subsurface reservoir because of the geothermal heat and hydrostatic pressure. In other words, there is a phase difference between the captured and stored CO_2, suggesting that a phase transition between the gas and liquid phases and the transition from subcritical to supercritical can occur in a CO_2 transportation and injection system. In transportation and injection systems operating under these conditions, even slight changes in the operating conditions and the external environmental conditions can lead to significant changes in the temperature and pressure of the CO_2. The changes in the temperature and pressure at the supercritical point and in a region near the pseudocritical line can lead to rapid changes in the properties, thus causing flow instability in the entire transportation and injection system. Therefore, in this study, the CO_2 flow instability was identified at the critical point and in the near-critical region during the CO_2 transportation and injection process, and its causes and effects were analyzed. It should be noted that this study has primarily concentrated on a numerical analysis of the CO_2 flow instability associated with the critical point and near-critical region. Nevertheless, experimental verification of the aforementioned unstable behavior would be helpful, and this will be considered in future research. The main conclusions from the results of this study are as follows.

(1)   The regions corresponding to the operating conditions causing flow instability were categorized into four areas; at the critical point, in the near-critical region, in a region near the pseudocritical line, and in the supercritical region. The pressure must be higher than 80 bar ($P > Ppc = 80$ bar) to prevent instability at temperatures lower than the pseudocritical temperature of ca. 34.5 °C ($T < Tpc = 34.5$ °C). Furthermore, the pressure must be at least 77 bar ($P > Ppc = 77$ bar) to avoid instability at temperatures higher than the pseudocritical temperature of ca. 34.5 °C ($T > Tpc = 34.5$ °C).

(2)   The density of CO_2 in the unstable region varies rapidly depending on the temperature and pressure. The rapid change in the density of CO_2 caused by the operating conditions in the unstable region caused an unstable flow with a trend similar to that of density wave oscillations resulting from the phase change in subcritical two-phase flow, even though the studied condition is a supercritical single-phase fluid.

(3) To describe the instability boundaries, the stability map was developed using the two dimensionless numbers which are the sub-pseudocritical number (N_{SUBPC}) and the trans-pseudo-critical number (N_{TPC}). The stability map revealed that the trends of the flow instability in CCS were similar with supercritical water.

(4) To verify whether the unstable region obtained in this study is specific only to the case of heating with the heater, additional numerical analysis was conducted on three horizontal pipelines. In all three cases, the flow instability of CO_2 was observed when the operating conditions entered the unstable region. Therefore, the unstable region of this study can be applied to the design of standard CO_2 pipelines. In other words, the results of this study can be used for safe and economic CO_2 transportation and the design of CO_2 pipelines.

Author Contributions: Conceptualization, C.H.; Simulation and Analysis, I.H.M. and C.H.; Funding Acquisition, S.-G.K.; Writing—Original Draft Preparation, I.H.M. and C.H.; Writing—Review & Editing, C.H.

Funding: This research was supported by a grant [KCG-01-2017-05] through the Disaster and Safety Management Institute funded by Korea Coast Guard of the Korean government and by the National Research Foundation of Korea (NRF) grant funded by the Korea government (Ministry of Science and ICT) (2017R1E1A1A03070672).

Acknowledgments: The authors would like to thank Schlumberger for their donation of OLGA.

Conflicts of Interest: The authors declare no conflict of interest.

References

1.   Metz, B.; Davidson, O.; de Coninck, H.; Loos, M.; Meyer, L. *IPCC Carbon Dioxide Capture and Storage*; Cambridge University Press: Cambridge, UK, 2005.

2.   Huh, C.; Kang, S.-G.; Park, M.-H.; Lee, K.-S.; Park, Y.-G.; Min, D.-J.; Lee, J.-S. Latest CO_2 transport, storage and monitoring R&D progress in Republic of Korea: Offshore geologic storage. *Energy Procedia* **2013**, *37*, 6520–6526.

3.   Min, I.H.; Huh, C. Development of new CO_2 heating process for offshore geological storage. *Int. J. Greenh. Gas Control* **2017**, *64*, 1–10. [CrossRef]

4.   Kang, K.; Seo, Y.; Chang, D.; Kang, S.-G.; Huh, C. Estimation of CO_2 transport costs in South Korea using a techno-economic model. *Energies* **2015**, *8*, 2176–2196. [CrossRef]

5.   Min, I.H.; Huh, C.; Choe, Y.S.; Kim, H.U.; Cho, M.I.; Kang, S.G. Numerical analysis of CO_2 behavior in the subsea pipeline, topside and wellbore with reservoir pressure increase over the injection period. *J. Korean Soc. Mar. Environ. Energy* **2016**, *19*, 286–296. [CrossRef]

6.   Pioro, I.; Mokry, S. Thermophysical properties at critical and supercritical conditions. In *Heat Trasnfer: Theoretical Analysis, Experimental Investigations and Industrial Systems*; Belmiloudi, A., Ed.; IntechOpen: Vienna, Austria, 2007; pp. 573–592.

7.   Lamorgese, A.; Ambrosini, W.; Mauri, R. Widom line prediction by the Soave–Redlich–Kwong and Peng–Robinson equations of state. *J. Supercrit. Fluids* **2018**, *133*, 367–371. [CrossRef]

8.   Lemmon, E.W.; Huber, M.L.; McLinden, M.O. *NIST Standard Reference Database 23: Reference Fluid Thermodynamic and Transport Properties-REFPROP*; version 9.1; NIST: Gaithersburg, MD, USA, 2013.

9.   Tian, R.; An, Q.; Zhai, H.; Shi, L. Performance analyses of transcritical organic Rankine cycles with large variations of the thermophysical properties in the pseudocritical region. *Appl. Therm. Eng.* **2016**, *101*, 183–190.

10. Cabeza, L.F.; de Gracia, A.; Fernández, A.I.; Farid, M.M. Supercritical CO_2 as heat transfer fluid: A review. *Appl. Therm. Eng.* **2017**, *125*, 799–810. [CrossRef]

11. Shoaib, A.M.; Bhran, A.A.; Awad, M.E.; El-Sayed, N.A.; Fathy, T. Optimum operating conditions for improving natural gas dew point and condensate throughput. *J. Nat. Gas Sci. Eng.* **2018**, *49*, 324–330. [CrossRef]

12. Guerrero-Zárate, D.; Estrada-Baltazar, A.; Iglesias-Silva, G.A. Calculation of critical points for natural gas mixtures with the GERG-2008 equation of state. *Fluid Phase Equilib.* **2017**, *437*, 69–82. [CrossRef]

13. Wang, T.; Ding, C.; Ding, G.; Zhao, D.; Ren, T. A fast calculation method of thermodynamic properties of variable-composition natural gas mixtures in the supercritical pressure region based on implicit curve-fitting. *J. Nat. Gas Sci. Eng.* **2017**, *43*, 96–109. [CrossRef]

14. Chaczykowski, M.; Zarodkiewicz, P. Simulation of natural gas quality distribution for pipeline systems. *Energy* **2017**, *134*, 681–698. [CrossRef]

15. Wang, P.; Yu, B.; Deng, Y.; Zhao, Y. Comparison study on the accuracy and efficiency of the four forms of hydraulic equation of a natural gas pipeline based on linearized solution. *J. Nat. Gas Sci. Eng.* **2015**, *22*, 235–244. [CrossRef]

16. Pambour, K.A.; Bolado-Lavin, R.; Dijkema, G.P.J. An integrated transient model for simulating the operation of natural gas transport systems. *J. Nat. Gas Sci. Eng.* **2016**, *28*, 672–690. [CrossRef]

17. Guandalini, G.; Colbertaldo, P.; Campanari, S. Dynamic modeling of natural gas quality within transport pipelines in presence of hydrogen injections. *Appl. Energy* **2017**, *185*, 1712–1723. [CrossRef]

18. Ma, Y.; Liu, M.; Yan, J.; Liu, J. Thermodynamic study of main compression intercooling effects on supercritical CO_2 recompression Brayton cycle. *Energy* **2017**, *140*, 746–756. [CrossRef]

19. Shao, L.L.; Zhang, C.L. Thermodynamic transition from subcritical to transcritical CO_2 cycle. *Int. J. Refrig.* **2016**, *64*, 123–129. [CrossRef]

20. Gupta, S.; Saltanov, E.; Mokry, S.J.; Pioro, I.; Trevani, L.; McGillivray, D. Developing empirical heat-transfer correlations for supercritical CO_2 flowing in vertical bare tubes. *Nucl. Eng. Des.* **2013**, *261*, 116–131. [CrossRef]

21. Han, S.H.; Chang, D.; Huh, C. Efficiency analysis of radiative slab heating in a walking-beam-type reheating furnace. *Energy* **2011**, *36*, 1265–1272. [CrossRef]

22. Chen, Q.; Zhao, S.Y.; Chen, X.; Li, X. The power flow topology of heat transfer systems at supercritical conditions for performance analysis and optimization. *Int. J. Heat Mass Transf.* **2018**, *118*, 316–326. [CrossRef]

23. Ma, T.; Chu, W.X.; Xu, X.Y.; Chen, Y.T.; Wang, Q.W. An experimental study on heat transfer between supercritical carbon dioxide and water near the pseudo-critical temperature in a double pipe heat exchanger. *Int. J. Heat Mass Transf.* **2016**, *93*, 379–387. [CrossRef]

24. Rao, N.T.; Oumer, A.N.; Jamaludin, U.K. State-of-the-art on flow and heat transfer characteristics of supercritical CO_2 in various channels. *J. Supercrit. Fluids* **2016**, *116*, 132–147. [CrossRef]

25. Archana, V.; Vaidya, A.M.; Vijayan, P.K. Flow transients in supercritical CO_2 natural circulation loop. *Procedia Eng.* **2015**, *127*, 1189–1196. [CrossRef]

26. Chen, L.; Zhang, X.R.; Yamaguchi, H.; Liu, Z.S. Effect of heat transfer on the instabilities and transitions of supercritical CO_2 flow in a natural circulation loop. *Int. J. Heat Mass Transf.* **2010**, *53*, 4101–4111. [CrossRef]

27. Swapnalee, B.T.; Vijayan, P.K.; Sharma, M.; Pilkhwal, D.S. Steady state flow and static instability of supercritical natural circulation loops. *Nucl. Eng. Des.* **2012**, *245*, 99–112. [CrossRef]

28. Liu, G.; Huang, Y.; Wang, J.; Lv, F.; Liu, S. Experimental research and theoretical analysis of flow instability in supercritical carbon dioxide natural circulation loop. *Appl. Energy* **2017**, *205*, 813–821. [CrossRef]

29. Chen, L.; Zhang, X.R.; Deng, B.L.; Jiang, B. Effects of inclination angle and operation parameters on supercritical CO_2 natural circulation loop. *Nucl. Eng. Des.* **2013**, *265*, 895–908. [CrossRef]

30. Zuber, N. *An Analysis of Thermally Induced Flow Oscillations in the Near-Critical and Supercritical Thermodynamic Region*; NASA: Huntsville, AL, USA, 1966.

31. Ambrosini, W. On the analogies in the dynamic behaviour of heated channels with boiling and supercritical fluids. *Nucl. Eng. Des.* **2007**, *237*, 1164–1174. [CrossRef]

32. Xiong, T.; Yan, X.; Xiao, Z.; Li, Y.; Huang, Y.; Yu, J. Experimental study on flow instability in parallel channels with supercritical water. *Ann. Nucl. Energy* **2012**, *48*, 60–67. [CrossRef]

33. KRISO. *Whole CCS System Operation Philosophy, Korea Clean Carbon Storage Project 2025 Pre-FEED Report*; KRISO: Dae-jeon, Korea, 2016.

34. Schlumberger OLGA Dynamic Multiphase Flow Simulator. Available online: http://www.software.slb.com/products/foundation/Pages/olga.aspx (accessed on 3 September 2014).

35. Huh, C.; Cho, M.I.; Kang, S.G. Numerical analysis on depressurization of high pressure carbon dioxide pipeline. *J. Korean Soc. Mar. Environ. Energy* **2016**, *19*, 52–61. [CrossRef]

36. KNOC. *Carbon Storage Potential in the Ulleung Basin, South Korea: Prospect Modelling and Ranking*; KNOC: Ulsan, Korea, 2015.

37. Dorao, C.A. Effect of inlet pressure and temperature on density wave oscillations in a horizontal channel. *Chem. Eng. Sci.* **2015**, *134*, 767–773. [CrossRef]
38. Su, Y.; Feng, J.; Zhao, H.; Tian, W.; Su, G.; Qiu, S. Theoretical study on the flow instability of supercritical water in the parallel channels. *Prog. Nucl. Energy* **2013**, *68*, 169–176. [CrossRef]
39. Ambrosini, W.; Sharabi, M. Dimensionless parameters in stability analysis of heated channels with fluids at supercritical pressures. *Nucl. Eng. Des.* **2008**, *238*, 1917–1929. [CrossRef]

Article

Anti-Agglomerator of Tetra-n-Butyl Ammonium Bromide Hydrate and Its Effect on Hydrate-Based CO_2 Capture

Rong Li [1,2,3], Xiao-Sen Li [1,2,4], Zhao-Yang Chen [1,2,4,*], Yu Zhang [1,2,4], Chun-Gang Xu [1,2,4] and Zhi-Ming Xia [1,2,4]

[1] Guangzhou Institute of Energy Conversion, Chinese Academy of Sciences, Guangzhou 510640, China; lirong@ms.giec.ac.cn (R.L.); lixs@ms.giec.ac.cn (X.-S.L.); zhangyu1@ms.giec.ac.cn (Y.Z.); xucg@ms.giec.ac.cn (C.-G.X.); xiazm@ms.giec.ac.cn (Z.-M.X.)

[2] Key Laboratory of Gas Hydrate, CAS and Guangzhou Center for Gas Hydrate Research, CAS, Guangzhou 510640, China

[3] Nano Sciences and Technology Institute, University of Science and Technology of China, Suzhou 215123, China

[4] Guangdong Key Laboratory of New and Renewable Energy Research and Development, Guangzhou 510640, China

* Correspondence: chenzy@ms.giec.ac.cn; Tel.: +86-20-8705-8468

Received: 14 December 2017; Accepted: 31 January 2018; Published: 8 February 2018

Abstract: Tetra-n-butyl ammonium bromide (TBAB) was widely used in the research fields of cold storage and CO_2 hydrate separation due to its high phase change latent heat and thermodynamic promotion for hydrate formation. Agglomeration always occurred in the process of TBAB hydrate generation, which led to the blockage in the pipeline and the separation apparatus. In this work, we screened out a kind of anti-agglomerant that can effectively solve the problem of TBAB hydrate agglomeration. The anti-agglomerant (AA) is composed of 90% cocamidopropyl dimethylamine and 10% glycerol, which can keep TBAB hydrate of 19.3–29.0 wt. % in a stable state of slurry over 72 h. The microscopic observation of the morphology of the TBAB hydrate particles showed that the addition of AA can greatly reduce the size of the TBAB hydrate particles. CO_2 gas separation experiments found that the addition of AA led to great improvement on gas storage capacity, CO_2 split fraction and separation factor, due to the increasing of contact area between gas phase and hydrate particles. The CO_2 split fraction and separation factor with AA addition reached up to 70.3% and 42.8%, respectively.

Keywords: CO_2 separation; TBAB; IGCC; anti-agglomerant; micromorphology; hydrate

1. Introduction

As a result of the excessive emission of CO_2 gas, global warming has caused extensive concern throughout the world. The Intergovernmental Panel on Climate Change (IPCC) report shows that by the end of the century, the content of CO_2 in the atmosphere will have rapidly increased to 575–950 ppm without control, and as a consequence, this will directly cause the global mean temperature to elevate 0.7–4.0 °C and will lead to catastrophic results [1]. At present, about one third of the global CO_2 emissions are from coal-fired power plants [2], and most of them are faced with the problem of low power efficiency and difficulties in CO_2 capture. The Integrated Gasification Combined Cycle Power Generation System (IGCC) is the most promising clean and efficient coal-fired process currently, which is proven to be used to gradually replace traditional coal technology. The syngas after coal gasification and reforming mainly contains 40 mol % CO_2 and 60 mol % H_2 [3], and its purification treatment (CO_2 removal) not only has an important influence on power generation efficiency, but also has practical significance for CO_2 emission reduction. Therefore, the development of targeted CO_2 separation and capture technology is vital for the development of the IGCC power generation system.

Traditional CO_2 separation technology mainly includes chemical absorption, physical adsorption, membrane separation, cryogenic distillation etc. However, these methods have their respective issues of high costs, easily causing secondary pollution and so forth in practical application [4]. Therefore, it is necessary to find a more efficient and environmentally friendly method for CO_2 separation. Hydrate separation technology is a new gas separation method that utilizes different gas hydrate phase equilibrium conditions to enrich the gas with lower phase equilibrium conditions in the hydrate phase, thereby achieving the goal of gas separation [5]. Since the outlet pressure of the IGCC syngas is between 2 and 7 MPa, the hydrate method does not require additional pressurization equipment to pressurize the syngas, thereby significantly reducing the operating costs. However, the industrial application of the technology still faces the problems of low hydrate formation rate, relatively high operating pressure and hydrate blockage in separating equipment. Previous studies have shown that the addition of tetrahydrofuran [6], propane [7], cyclopentane [8] and other thermodynamic promoters can greatly moderate the hydrate formation conditions and accelerate the hydrate formation. However, these promoters are faced with the problem of volatilizing in the practical application.

Tetrabutylammonium bromide (TBAB) has the characteristic of being non-volatile in aqueous solutions, which has been shown to greatly moderate the formation conditions of gas hydrates [9]. However, during the course of the experiment, it was found that TBAB hydrate is highly prone to coalescence and deposit, and can cause blockage of hydrate separation equipment extremely easily. The addition of anti-agglomerant to the hydrate system to prevent hydrate agglomeration is considered to be an effective way to improve the hydrate slurry flow [10,11]. The anti-agglomerant is usually a kind of surfactant that does not affect the nucleation and growth of hydrates but prevents hydrate particle agglomeration [12]. In the 1980s, the French Petroleum Institute confirmed that nonionic amphiphilic surfactants can effectively inhibit gas hydrate from generating [13–15]. Subsequently, Shell found that the quaternary ammonium surfactant was an ideal agglomeration inhibitor and argued that two or more n-butyl, n-pentyl or isopentyl ammonium salts had a good performance of delaying hydrate crystal growth [16]. In addition, the BJ UnichemCompany found that quaternized polyamine polyethers are more efficient than polyvinyl caprolactam (PVCap) and some quaternary anti-agglomerants in preventing the accumulation of tetrahydrofuran hydrate [16,17]. However, most of these anti-agglomerants are used for preventing CH_4 hydrate blockage and must be used in high-ratio oil/water systems (70 vol %), which tend to produce grease in the pipeline. If the low water content system is used for CO_2 hydrate separation, the amount of hydrate formed will greatly reduce, thus reducing the gas separation capacity greatly.

Up to now, little attention has been devoted to anti-agglomeration of the oil-free hydrate system. In this work, we investigated the effect of several different anti-agglomerants on the aggregation of TBAB hydrate particles in oil-free systems and screened the best one. Then we investigated its effect on carbon dioxide separation from the IGCC syngas under different conditions, and, combined with the microscopic analysis method to explore the anti-agglomeration mechanism of the anti-agglomerant.

2. Results and Discussion

2.1. Anti-Agglomerant Screening

The results of eleven hydrate anti-agglomeration experiments are listed in Table 1, and the photographs of TBAB hydrate with different anti-agglomerants are shown in Figure 1. After the inert gas bubbling, the TBAB hydrate generated rapidly in one minute, and the hydrates exhibited a state of slurry. After half an hour, only the hydrates in the tubes with Span80 and cocamidopropyl dimethylamine and 10% glycerol (AA) addition still exhibited slurry state, and another an hour later, only the hydrate in the tube with AA addition kept the state of slurry. The hydrates with the other anti-agglomerants addition agglomerated into block solid in 1.5 h. Thus, AA is determined to be the most effective anti-agglomerant.

Table 1. Results of anti-agglomerant screening.

No.	Anti-Agglomerant	$\frac{TBAB^{a}}{wt.\%}$	$\frac{AA^{b}}{wt.\%}$	$\frac{V^{c}}{mL}$	$\frac{T^{d}}{K}$	$\frac{t^{e}}{h}$	Result [f]
1	HPMA [g]	0.5	16.2	30	275.15	1.5	No
2	HPMA	3.0	16.2	30	275.15	1.5	No
3	AA [h]	0.5	16.2	30	275.15	1.5	Yes
4	AA	3.0	16.2	30	275.15	1.5	Yes
5	Span80	0.5	16.2	30	275.15	1.5	No
6	Span80	3.0	16.2	30	275.15	1.5	No
7	PAM [i]	0.5	16.2	30	275.15	1.5	No
8	PAM	3.0	16.2	30	275.15	1.5	No
9	SDS [j]	0.5	16.2	30	275.15	1.5	No
10	SDS	3.0	16.2	30	275.15	1.5	No
11	-	0	16.2	30	275.15	1.5	No

[a] Mass concentration of anti-agglomerant, with accuracy of ±0.01 g. [b] Mass concentration of tetra-n-butyl ammonium bromide, with accuracy of ±0.01 g. [c] Volume of tetra-n-butyl ammonium bromide aqueous solution, with accuracy of ±0.1 mL. [d] Experimental temperature, with accuracy of ±0.1 K. [e] Duration of experiment. [f] Yes refers to anti-agglomeration, no refers to aggregation. [g] Polymaleic acid. [h] 90% cocamidopropyl dimethylamine +10% glycerol. [i] Polyacrylamide. [j] Sodium dodecyl sulfate.

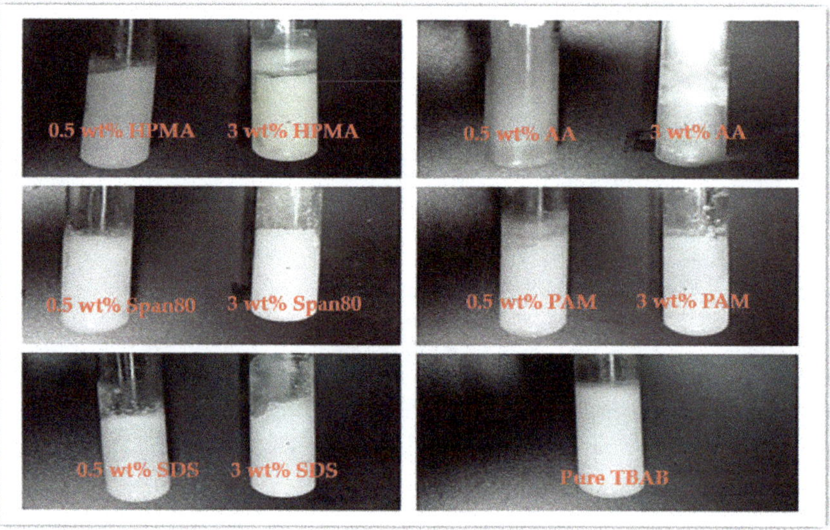

Figure 1. Photograph of Tetrabutylammonium bromide (TBAB) hydrate with different anti-agglomerants after 1.5 h.

2.2. Effect of Anti-Agglomerant Dosage on Its Anti-Agglomeration Duration

The stable time of hydrate slurry is an important parameter for assessing the performance of anti-agglomerant. The photographs of TBAB hydrate with different amounts of AA addition are displayed in Figure 2. The TBAB concentration is 16.2 wt. %. The optimal anti-agglomerant concentration was determined by measuring the stable time of the hydrate slurry. As shown in Figure 2, after 24 h, the hydrate with 0.5 wt. % AA agglomerated, while the hydrates with 3.0 wt. % AA still maintained the state of slurry. However, another 12 h later, the hydrate slurry with 3.0 wt. % AA addition also exhibited agglomeration. As the AA dosage increased to 4 wt. %, the TBAB hydrate slurry remained anti-agglomeration up to 72 h. This indicates that the anti-agglomeration duration of AA is positively correlated with its concentration used. The minimum AA dosage required for 16.2 wt. % TBAB solution to maintain slurry state in 72 h is about 4 wt. %.

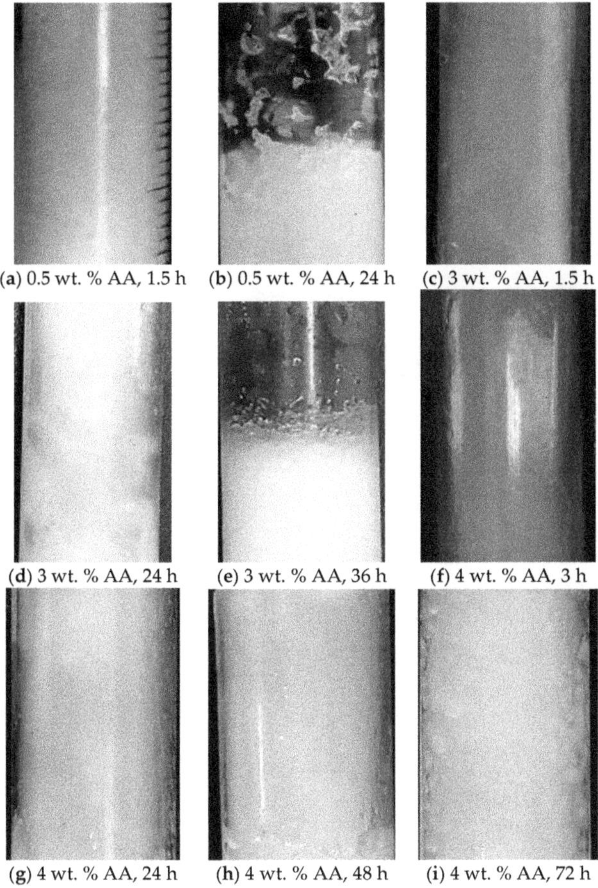

(a) 0.5 wt. % AA, 1.5 h   (b) 0.5 wt. % AA, 24 h   (c) 3 wt. % AA, 1.5 h

(d) 3 wt. % AA, 24 h   (e) 3 wt. % AA, 36 h   (f) 4 wt. % AA, 3 h

(g) 4 wt. % AA, 24 h   (h) 4 wt. % AA, 48 h   (i) 4 wt. % AA, 72 h

Figure 2. The photographs of TBAB hydrate with different amount of 90% cocamidopropyl dimethylamine +10% glycerol (AA) addition.

2.3. Effect of TBAB Concentration on AA's Critical Concentration Needed

In this work, the minimum AA concentration needed to prevent TBAB hydrate from agglomerating within 72 h is defined as AA's critical concentration. The stepwise method to increase AA concentration was used to determine the anti-agglomerant critical concentration in our experiments. Figure 3 shows the AA's critical concentration for TBAB aqueous solutions with different concentrations. The AA's critical concentration increases with TBAB concentration when TBAB concentration is lower than 26.0 wt. %, and then levels off when TBAB concentration increases further.

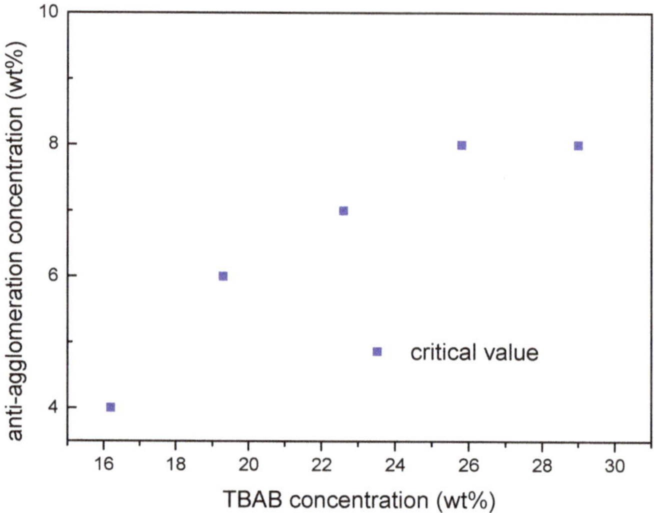

Figure 3. AA's critical minimum concentration for different TBAB solutions.

2.4. Micromorphology of TBAB Hydrate Slurry

The micrographs of TBAB hydrate with and without AA addition are shown in Figure 4. The photos on the right-hand side are the enlarged view of the red circle in the photos of the left side. Figure 4a shows the micromorphology of hydrate crystals generated in the TBAB system. It could be found that the pure TBAB hydrate particles are rod-like crystal of relatively large size that stick together and grows in clusters. Figure 4b shows the morphology of hydrate formation in the TBAB/AA system; it can be seen that the TBAB hydrate particles with small size are dispersed in the solution. This indicates that the anti-agglomerant played an important role in preventing TBAB hydrate particles from aggregation. However, the anti-agglomeration mechanism of AA still needs further study.

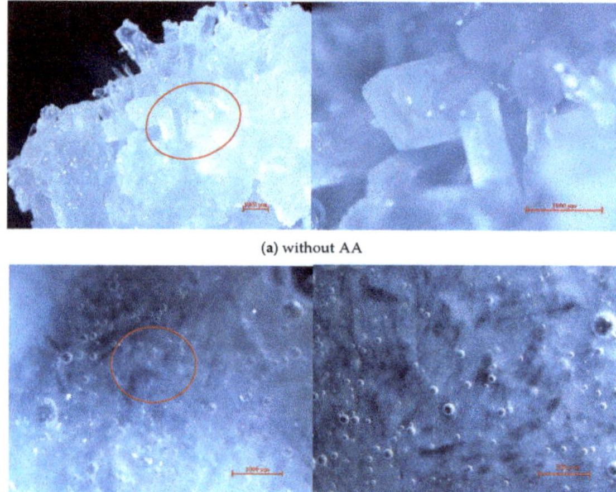

Figure 4. Micrographs of TBAB hydrate.

2.5. Analysis of the AA's Anti-Agglomeration Mechanisms

In the past, the formation of water-in-oil emulsions has been proposed as a requirement for hydrate anti-agglomeration [11,16,18–20]. However, the AA proposed in this work is effective in an oil-free system. The anti-agglomeration mechanisms of AA are schematically shown in Figure 5 [21]. The nonionic surfactant cocamidopropyl dimethylamine forms micelles in the aqueous solution before hydrate generation. Because the hydrogen bonds between amino groups and water molecules is stronger than hydrogen bonds between water molecules, the cocamidopropyl dimethylamine can form hydrogen bonds with the water molecules on the surface of the hydrates. Once hydrates form in the solution, some of the micelles dissociate and the cocamidopropyl dimethylamine molecules adsorb on the surface of hydrate crystal nucleus. Meanwhile, the hydrophobic tail of the surfactant will cover the surface of the hydrate particles because the dielectric constant of TBAB hydrates is less than that of water [21], thus preventing the hydrates from growth and further agglomeration.

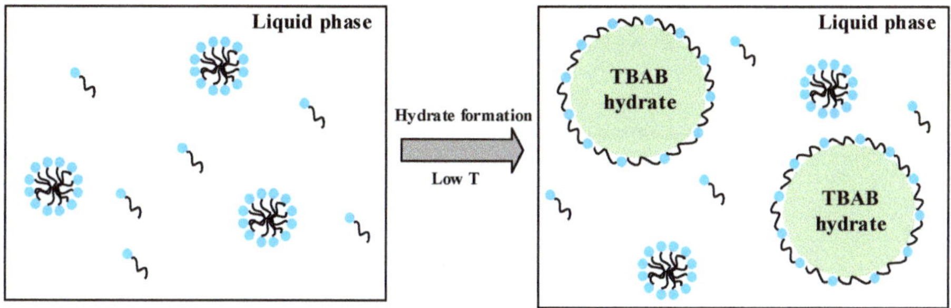

Figure 5. Anti-agglomeration mechanisms of AA.

2.6. Effect of AA's Anti-Agglomeration on CO_2 Separation

In order to study the effect of AA addition on CO_2 separation efficiency, six groups of comparative experiments were conducted at the initial conditions of 277.15 K and 4.000 MPa. The gas-liquid ratio was 210 vol %/125 vol %. The experiment conditions and results are listed in Table 2. As can be seen from Table 2, the CO_2 concentration in residual gas decreased from 30.62% (in TBAB system) to 17.33% (in TBAB/AA binary system) with AA addition.

Figures 6 and 7 display the gas storage capacity, CO_2 split fraction and CO_2 separation factor under different separating conditions, respectively.

Table 2. Experimental conditions and results for simulated syngas hydrate separation with different additives.

Experiment No.	$\dfrac{TBAB^a}{wt.\%}$	$\dfrac{AA^b}{wt.\%}$	$\dfrac{T^c}{K}$	$\dfrac{P^d}{MPa}$	$\dfrac{L/G}{VOL\%/VOL\%}$	$\dfrac{X_G{}^e}{\%}$	S.Fr.^{f}	S.F.^{g}
1	19.3	0	277.15	4.000	125/210	30.62	0.346	14.4
2	19.3	6	277.15	4.000	125/210	17.33	0.703	42.8
3	22.6	0	277.15	4.000	125/210	29.81	0.401	6.2
4	22.6	7	277.15	4.000	125/210	19.97	0.648	29.3
5	25.8	0	277.15	4.000	125/210	20.25	0.647	26.2
6	25.8	8	277.15	4.000	125/210	17.96	0.696	28.3

[a] Mass concentration of tetra-n-butyl ammonium bromide, with accuracy of ±0.01 g. [b] Mass concentration of anti-agglomerant, with accuracy of ±0.01 g. [c] Experimental temperature, with accuracy of ±0.1 K. [d] Experimental initial pressure, with accuracy of ±0.02 MPa. [e] Concentration of CO_2 in the residual gas phase, with accuracy of ±0.1%. [f] Split fraction of CO_2. [g] Separation factor.

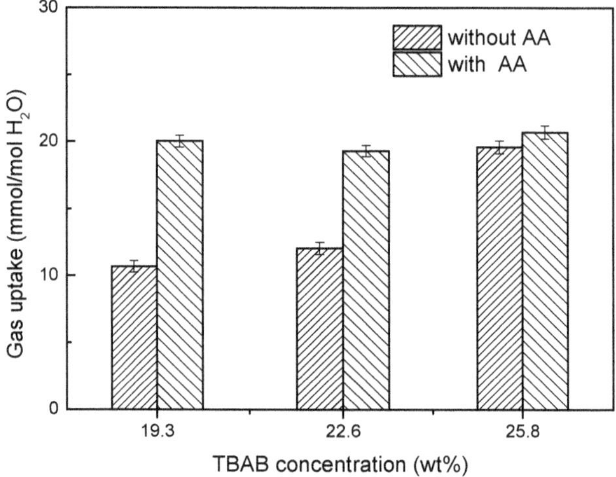

Figure 6. Gas storage capacity in different conditions.

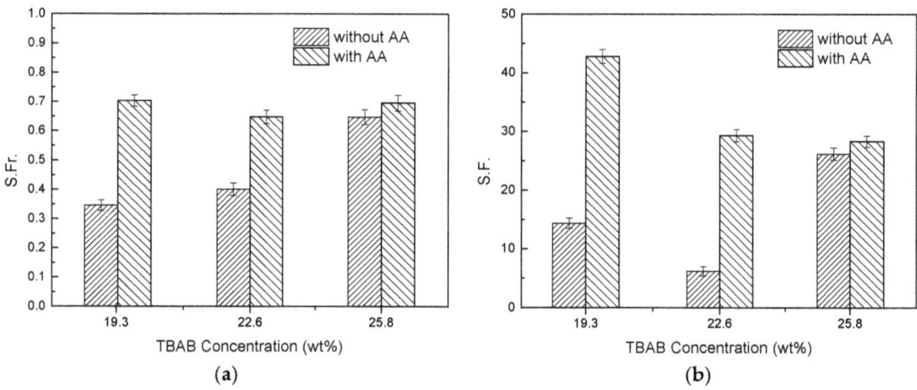

Figure 7. CO_2 split fraction (**a**) and separation factor (**b**) in different conditions.

As can be seen from Figure 6, for the same TBAB concentration, the gas storage capacity of hydrate formation in the presence of AA was much higher than that without AA. Taking 19.3 wt. % TBAB as an example, relative to pure a TBAB system, the gas storage capacity of hydrate formation in the TBAB/AA binary system increased by 87.8% (from 10.66 mmol/mol H_2O to 20.02 mmol/mol H_2O). This is because TBAB hydrate particles would agglomerate and adhere together in the process of hydrate formation without AA addition, thus preventing the CO_2 from getting into the TBAB hydrate cage. As for TBAB/AA binary system with the same TBAB concentration, the presence of AA can prevent TBAB hydrate aggregation, thus can increase the contact area between CO_2 and TBAB hydrate particles, so that more CO_2 gas can get into TBAB cages. For a pure TBAB system, the gas storage capacity increased with TBAB concentration. However, for the TBAB/AA binary system, the gas storage capacity was approximately equal in three different TBAB/AA binary systems, which means the synergistic effect of AA and TBAB on hydrate-based CO_2 separation was weakened with the increase in TBAB concentration. This is because the amount of TBAB hydrate cages generated in the solution increased with TBAB concentration in the pure TBAB system, so that more CO_2 gas can get into the cages even if hydrate particles adhere to each other. TBAB hydrate particles would disperse in the solution with AA addition, thus increases the contact area between gas phase and hydrate particles, which will lead to an increase of gas transfer. Actually, because only 5^{12} cavity of TBAB hydrate can accommodate gas molecules, and each TBAB hydrate has three 5^{12} cavity, 1 mole TBAB hydrate can capture 3 mol CO_2. The amount of CO_2 in the feed gas was about 1.55 mol, thus the amount of CO_2 relative to the amount of TBAB cages was insufficient in the three concentrations (as shown in Table 3). Therefore, after achieving phase equilibrium, the gas storage capacity was approximately equal in three different TBAB/AA binary systems even if the TBAB hydrate particles dispersed in the solution.

Table 3. Theoretical value of CO_2 storage in different conditions.

C_{TBAB}/wt. %	V_{TBAB}/mL	n_{TBAB}/mol	$n_{1_{CO_2}}$ [a]/mol	$n_{2_{CO_2}}$ [b]/mol
19.3	125	0.075	0.225	0.155
22.6	125	0.088	0.264	0.155
25.8	125	0.100	0.300	0.155

[a] Theoretical value of CO_2 storage. [b] Mole number of CO_2 in feed gas.

As can be seen from Figure 7, the CO_2 split fractions of hydrate formation in TBAB/AA binary system are higher than that in TBAB system without AA. The CO_2 separation factor of hydrate-based CO_2 separation increased significantly in the TBAB solution containing AA. On the basis of these, it could be concluded that the addition of AA has positive effect on hydrate-based CO_2 separation in TBAB solution.

The comparison of CO_2 separation results using TBAB as the promoter is shown in Table 4. The CO_2 spilt fraction of TBAB/AA system obtained in this work increased by 49.6% compared to the highest value reported in the literature [22]. This is mainly because the concentration of TBAB used in this work was much higher than that in Gholinezhad et al. [22], and the addition of AA can effectively disperse TBAB hydrate particles, so more CO_2 can get into TBAB cages.

Table 4. Comparison of CO_2 separation results using TBAB as the promoter.

References	TBAB [a]	$\frac{T^b}{K}$	$\frac{P^c}{MPa}$	$\frac{L/G}{VOL\%/VOL\%}$	S.Fr.	S.F.
Gholinezhad et al. [22]	5 and 10 wt. %	273.5–273.9	3.8–3.9	50/50	41–47%	15.7–28.0
XU et al. [23]	0.29 mol %	274.15	3.0	180/120	14.5%	-
Kim et al. [24]	0.5, 1.0 and 3.0 mol %	284.15–287.15	3.0	-	10.1–24.1%	16.1–26.0
This work	19.3, 22.6 and 25.8 wt. %	277.15	4.0	125/210	34.6–64.7%	6.2–26.2
This work	19.3, 22.6 and 25.8 wt. % with AA	277.15	4.0	125/210	64.8–70.3%	28.3–42.8

[a] Tetra-n-butyl ammonium bromide. [b] Experimental temperature, with accuracy of ± 0.1 K. [c] Experimental initial pressure, with accuracy of ± 0.02 MPa.

3. Materials and Methods

3.1. Apparatus

Figure 8a shows a schematic diagram of CO_2 separation apparatus used for hydrate kinetic study. The experimental equipment mainly consists of a cylindrical crystallizer (CR) and a balance tank (SV), which are immersed in a water bath for temperature control. CR and SV are made of high-pressure stainless steel, so as to withstand pressure up to 25 MPa, and the inner volume is 336 mL and 1350 mL, respectively. The temperature and pressure of CR and SV are measured by a Pt100 temperature sensor (JM6081, Hefei Heavy Tripod Machinery and Electrical Equipment Co., Ltd., Hefei, China) with an accuracy of ± 0.1 K and a pressure sensor (Model 552, Boxborough, MA, USA) with an accuracy of ± 0.02 MPa, respectively. All the temperature and pressure data are collected and stored in a computer (PC). There are two visualization windows on the front and back of the CR to observe the process of hydrate formation. The CR is equipped with an electromagnetic stirrer (400 r/min, Chengbang Science and Technology Co., Ltd., Sichuan, China) for accelerating the gas and liquid mixing. The pressure of CR can be maintained at a stable value through the PID control valve (E-ATR-7/250/I, Atos Co., Ltd., Sesto Calende, Italy) between CR and SV. The composition of residual gas in the CR after the hydrate formation and dissociation were analyzed by gas chromatography (GC522, Shanghai Wufeng Scientific Instrument Co., Ltd., Shanghai, China).

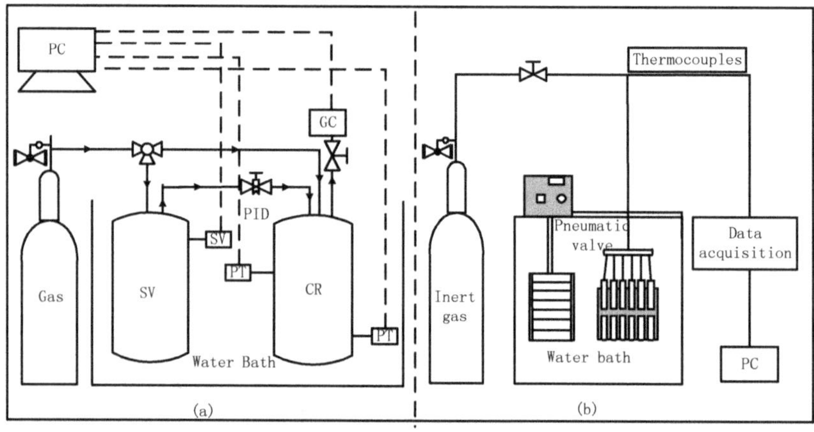

Figure 8. Schematic diagram of experimental equipment. (**a**) CO_2 separation apparatus; (**b**) Anti-agglomerant screening device.

The anti-agglomerant screening device diagram is shown in Figure 8b. A series of tubular reactors are placed in a low-temperature water bath (Model XT5218-B12, Xutemp Temptech Co., Ltd., Hangzhou, China) for temperature control, with an accuracy of ±0.1 K and range from −25 to 90 °C. The inert gas was introduced to the bottom of the tubular reactor through the pressure-reducing valve (115H-200, Gentec (Shanghai) Corporation, Shanghai, China) and pneumatic valve (PK4, Changzhou Chvird Automation Equipment Co., Ltd., Changzhou, China) to agitate the solution by bubbling. The temperatures of the aqueous solution in the tubular reactors were measured by temperature sensors, and collected by the Agilent Data Acquirement System (34970A, Agilent technologies Co., Ltd., China).

3.2. Materials

All aqueous solutions were prepared by the deionized water (with 18.25 MΩ/cm resistivity) produced by an ultra-pure water system (GREEN-10T, Nanjing EPED Science and Technology Development Co., Ltd., Nanjing, China). All chemical reagents used in this work are analytically pure, and are listed in Table 5.

Table 5. The chemical reagents used in this work.

No.	Name	Purity	Supplier
1	TBAB	99.0%	Aladdin Industrial Corporation (Shanghai agent, China)
2	Sodium dodecyl sulfate (SDS)	analytical pure	Tianjin Fuchen Chemical Reagent Factory
3	Span80	99.0%	Tianjin Kermel Chemical Reagent Factory Co., Ltd.
4	Polyacrylamide (PAM)	MW. 5 million	Shanghai Macklin Biochemicals Co., Ltd.
5	Polymaleic acid	50 vol %	Shanghai Macklin Biochemicals Co., Ltd.
6	Cocamidopropyl dimethylamine		Shanghai Yincong New Material Technology Co., Ltd.
7	Glycerin	99.0%	MYM Biological Technology Co., Ltd.
8	Simulated syngas	39.98 mol % CO_2 + 60.02 mol % H_2	Huate Gas Co., Ltd., China.

Five kinds of polymerization inhibitors used in this work are polymaleic acid (HPMA), AA (90% cocamidopropyl dimethylamine + 10% glycerol), Span80, polyacrylamide (PAM) and sodium dodecyl sulfate (SDS). Polymaleic acid has good scale inhibition performance and can also be used to improve the dispersion of particles. The anti-agglomerant named AA in this work consisted of 90% cocamidopropyl dimethylamine and 10% glycerol. Cocamidopropyl dimethylamine is a kind of nonionic surfactant, and Sun et al. [21] found the surfactant can effectively prevent methane hydrate agglomeration in oil-free systems. Glycerol is a kind of hydrophilic substance, thus can play the role of co-solvent and increase the solubility of cocamidopropyl dimethylamine in water. Span 80 is a kind of lipophilic nonionic surfactant with an HLB of 4.3 and can prevent the agglomeration of tetrahydrofuran hydrate [25]. Polyacrylamide is a kind of water-soluble polymer compound, which can effectively reduce the frictional resistance of the fluid. Sodium dodecyl sulfate is highly hydrophilic and can effectively reduce the surface tension of the solution.

3.3. Methods

3.3.1. Anti-Agglomerant Screening Experiments

Eleven tubular reactors were used for anti-agglomerant screening. Each reactor contained 30 mL 0.5 mol/L TBAB aqueous solution and different concentration of anti-agglomerant. The temperature of the water bath was set as 275.15 K. Once the solution temperatures in these reactors reached a steady state, the inert gas bubbling started to agitate the solutions. After the hydrate formation, the photograph of hydrate slurry was taken to evaluate the anti-agglomeration performance of the anti-agglomerant.

3.3.2. Observation of Micromorphology of TBAB Hydrate

The hydrate samples generated from pure TBAB solution and TBAB/AA binary system were placed under the electron microscopy (Nikon SMZ1500, Nikon Corporation, Tokyo, Japan), and adjusted to the appropriate focal length and magnification, then pictures for recording were taken.

3.3.3. CO_2 Hydrate Separation Experiments

For every experimental run, a cylindrical crystallizer (CR) was first filled with 125 mL aqueous solution with additive and pressurized with the simulated syngas to 4.000 MPa. Once CR was stabilized at 277.15 K, the agitator was turned on and the hydrate formation was awaited, and all the experimental data began to be recorded with time. After the hydrate formation finished, the residual gas phase in CR was sampled and analyzed by GC. Subsequently, the residual gas was quickly drained out of CR. After that, the temperature in CR was elevated to 293.15 K to make the hydrate dissociate completely. Finally, the composition of the gas phase from hydrate decomposition was analyzed by GC, each gas phase sample was measured three times and the test results averaged.

According to the mass conservation law, the total amount of gas is composed of the gas dissolved in the water, the gas encaged in the hydrate and the gas remaining in the residual gas phase. The amount of gas consumed during the hydrate formation (Δn_H) includes the gas dissolved in the water and the gas encaged in the hydrate, which can be calculated as follows [5]:

$$n_{G,0} = \Delta n_H + n_{G,t} \tag{1}$$

$$\Delta n_H = \left(\frac{PV}{zRT}\right)_{G,0} - \left(\frac{PV}{zRT}\right)_{G,t} \tag{2}$$

where z is the compressibility factor calculated by Pitzer's correlation [26], subscript t refers to time, subscript 0 refers to the initial time, subscript G refers to the gas phase in the CR.

The split fraction and separation factor can be evaluated according to the following equations [27]:

$$S.Fr. = \frac{n_{CO_2}^{H}}{n_{CO_2}^{Feed}}$$

(3)

$$S.F. = \frac{n_{CO_2}^{H} \times n_{H_2}^{G}}{n_{CO_2}^{G} \times n_{H_2}^{H}}$$

(4)

where *S.Fr.* and *S.F.* refer to separation fraction and separation factor, respectively. $n_{CO_2}^{Feed}$, $n_{CO_2}^{H}$ and $n_{CO_2}^{G}$ defined as the mole number of CO_2 in feed gas, in hydrate phase and in residual gas, respectively. $n_{H_2}^{G}$ and $n_{H_2}^{H}$ is the mole number of H_2 in residual gas and in hydrate phase, respectively.

4. Conclusions

In this work, we screened out a kind of anti-agglomerant AA (90% cocamidopropyl dimethylamine +10% glycerol) that can prevent TBAB hydrate from agglomerating effectively. The anti-agglomerating performance and mechanisms of AA were studied by TBAB hydrate formation experiment and microscopic observation by electron microscope. The effect of AA anti-agglomerant addition on CO_2 hydrate separation was studied by IGCC syngas separating experiments using TBAB as the hydrate formation promoter. The conclusions obtained are as follows:

1. The minimum AA dosage required increases with TBAB concentration in solution. A 6–8 wt. % AA addition can effectively prevent hydrate particles agglomeration in 19.3–29 wt. % TBAB aqueous solution. The TBAB hydrates slurry maintains stable over 72 h.
2. The micromorphology of hydrate particles shows that the addition of AA reduces the particle size of TBAB hydrate markedly because the cocamidopropyl dimethylamine molecules absorbs on the surface of TBAB hydrate crystal nucleus and forms a steric hindrance in the process of hydrate particle growth and agglomeration.
3. The IGCC syngas hydrate separating results indicate that AA addition can not only effectively prevent TBAB-CO_2 hydrate agglomeration and the blockage of the separating equipment, but also improve the gas transfer in the phase of the hydrate slurry, thus increase the hydrate separation efficiency drastically.
4. Compared to the separating results without AA addition, the CO_2 concentration in residual gas decreased from 30.62% to 17.33% with AA addition. Gas storage capacity, CO_2 split fraction and separation factor increased significantly, and CO_2 split fraction and separation factor reached 70.3% and 42.8%, respectively.

Acknowledgments: This work was supported by the grants from the National Natural Science Foundation of China (51576202, 51476174, 51736009), International S&T Cooperation Program of China (No.2015DFA61790) and National Key Research and Development Plan of China (No.2016YFC0304002), which are gratefully acknowledged.

Author Contributions: All the authors contributed to publish this paper. Rong Li, Xiao-Sen Li and Zhao-Yang Chen conceived and designed the experiments; Rong Li, Zhao-Yang Chen and Chun-Gang Xu performed the experiments and wrote the paper; Yu Zhang and Zhi-Ming Xia prepared for the experiments and analyzed the data.

Conflicts of Interest: The authors declare no conflict of interest.

References and Notes

1. Wigley, T.M.L.; Jain, A.K.; Joos, F.; Nyenzi, B.S.; Shukla, P.R. *Implications of Proposed CO₂ Emissions Limitations*; IPCC Technical Paper 4; Intergovernmental Panel on Climate Change: Geneva, Switzerland, 1997.
2. Freund, P.; Ormerod, W.G. Progress toward storage of carbon dioxide. *Energy Convers. Manag.* **1997**, *38*, S199–S204. [CrossRef]

3. Hendricks, C.A.; Blok, K.; Turkenburg, W.C. Technology and cost of recovering and storing carbon-dioxide from an integratedgasifier, combined-cycle plant. *Energy* **1991**, *16*, 1277–1293. [CrossRef]

4. Duc, N.H.; Chauvy, F.; Herri, J.M. CO_2 capture by hydrate crystallization—A potential solution for gas emission of steelmaking industry. *Energy Convers. Manag.* **2007**, *48*, 1313. [CrossRef]

5. Linga, P.; Kumar, R.N.; Englezos, P. Gas hydrate formation from hydrogen/carbon dioxide and nitrogen/carbon dioxide gas mixtures. *Chem. Eng. Sci.* **2007**, *62*, 4268–4276. [CrossRef]

6. Lee, H.J.; Lee, J.D.; Linga, P.; Englezos, P.; Kim, Y.S.; Lee, M.S.; Kim, Y.D. Gas hydrate formation process for pre-combustion capture of carbon dioxide. *Energy* **2009**, *35*, 2729–2733. [CrossRef]

7. Kumar, R.; Linga, P.; Ripmeester, J.A.; Englezos, P. Two-stage clathrate hydrate/membrane process for precombustion capture of carbon dioxide and hydrogen. *J. Environ. Eng. ASCE* **2009**, *135*, 411–417. [CrossRef]

8. Zhang, J.S.; Yedlapalli, P.; Lee, J.W. Thermodynamic analysis of hydrate-based pre-combustion capture of CO_2. *Chem. Eng. Sci.* **2009**, *64*, 4732–4736. [CrossRef]

9. Kang, S.P.; Lee, H. Recovery of CO_2 from flue gas using gas hydrate: Thermodynamic verification through phase equilibrium measurements. *Environ. Sci. Technol.* **2000**, *34*, 4397–4400. [CrossRef]

10. Mokhatab, S.; Wilkens, R.J.; Leontaritis, K.J. A review of strategies for solving gas-hydrate problems in subsea pipelines. *Energy Source Part A* **2007**, *29*, 39–45. [CrossRef]

11. Huo, Z.; Freer, E.; Lamar, M.; Sannigrahi, B.; Knauss, D.M.; Sloan, E.D. Hydrate plug prevention by anti-agglomeration. *Chem. Eng. Sci.* **2001**, *56*, 4979–4991. [CrossRef]

12. Huangjing, Z.; Minwei, S.; Abbas, F. Anti-agglomeration of natural gas hydrates in liquid condensate and crude oil at constant pressure conditions. *Fuel* **2016**, *180*, 187–193.

13. Sugier, A.; Bourgmayer, P.; Behar, E.; Freund, E. EP 323307, 5 July 1989.

14. Sugier, A.; Bourgmayer, P.; Behar, E.; Freund, E. EP 323774, 12 July 1989.

15. Sugier, A.; Bourgmayer, P.; Stern, R. EP 323775, 12 July 1989.

16. Zanota, M.L.; Dicharry, C.; Graciaa, A. Hydrate plug prevention by quaternary ammonium salts. *Energy Fuels* **2005**, *19*, 584–590. [CrossRef]

17. Kelland, M.A.; Svartaas, T.M.; Ovsthus, J.; Tomita, T.; Chosa, J. Studies on some zwit-terionic surfactant gas hydrate anti-agglomerants. *Chem. Eng. Sci.* **2006**, *61*, 4048–4059. [CrossRef]

18. Sloan, E.D. A changing hydrate paradigm—From apprehension to avoidance to risk management. *Fluid Phase Equilib.* **2005**, *228*, 67–74. [CrossRef]

19. Mehta, A.P.; Hebert, P.B.; Cadena, E.R.; Weatherman, J.P. Fulfilling the promise of low-dosage hydrate inhibitors: Journey from academic curiosity to successful field implementation. *SPE Prod. Facil.* **2003**, *18*, 73–79. [CrossRef]

20. Sloan, E.D.; Koh, C.A. *Clathrate Hydrates of Natural Gases*; CRC: Boca Raton, FL, USA, 2008.

21. Sun, M.; Firoozabadi, A. New surfactant for hydrate anti-agglomeration in hydrocarbon flowlines and seabed oil capture. *J. Colloid. Interface Sci.* **2013**, *402*, 312–319. [CrossRef] [PubMed]

22. Jebraeel, G.; Antonin, C.; Bahman, T. Separation and capture of carbon dioxide from CO_2/H_2 syngas mixture using semi-clathrate hydrates. *Chem. Eng. Res. Des.* **2011**, *89*, 1747–1751.

23. Xu, C.; Zhang, S.; Cai, J.; Chen, Z.; Li, X. CO_2 (carbon dioxide) separation from CO_2-H_2 (hydrogen) gas mixtures by gas hydrates in TBAB (tetra-n-butyl ammonium bromide) solution and Raman spectroscopic analysis. *Energy* **2013**, *59*, 719–725. [CrossRef]

24. Soo, M.K.; Ju, D.L.; Hyun, J.L.; Eun, K.L.; Yangdo, K. Gas hydrate formation method to capture the carbon dioxide for pre-combustion process in IGCC plant. *Int. J. Hydrog. Energy* **2011**, *36*, 1115–1121.

25. Taylor, C.J.; Dieker, L.E.; Miller, K.T.; Koh, C.A.; Sloan, E.D., Jr. Micromechanical adhesion force measurements between tetrahyfrofuran hydrate particles. *J. Colloid Interface Sci.* **2007**, *306*, 255–261. [CrossRef] [PubMed]

26. Smith, J.M.; Van Ness, H.C.; Abbott, M.M. *Introduction to Chemical Engineering Thermodynamics*; McGraw-Hill: New York, NY, USA, 2001.

27. Praveen, L.; Rajnish, K.; Peter, E. The clathrate hydrate process for post and pre-combustion capture of carbon dioxide. *J. Hazard. Mater.* **2007**, *149*, 625–629.

MDPI

St. Alban-Anlage 66

4052 Basel

Switzerland

Tel. +41 61 683 77 34

Fax +41 61 302 89 18

www.mdpi.com

Energies Editorial Office

E-mail: energies@mdpi.com

www.mdpi.com/journal/energies

www.ingramcontent.com/pod-product-compliance
Lightning Source LLC
LaVergne TN
LVHW071519180726
843512LV00014B/1119